CISM International Centre for Mechanical Sciences

Courses and Lectures

Volume 616

International Centre for Mechanical Sciences

For more than 40 years the book series edited by CISM, "International Centre for Mechanical Sciences: Courses and Lectures", has presented groundbreaking developments in mechanics and computational engineering methods. It covers such fields as solid and fluid mechanics, mechanics of materials, micro- and nanomechanics, biomechanics, and mechatronics. The papers are written by international authorities in the field. The books are at graduate level but may include some introductory material.

Valerio Cozzani · Fabrizio Paolacci
Editors

Management Strategies and Resilience Against NaTech Events

Technological Accidents Caused by Natural Hazards

Springer

Editors
Valerio Cozzani
DICAM—LISES, Laboratory of Industrial Safety and Environmental Sustainability
University of Bologna
Bologna, Italy

Fabrizio Paolacci
Department of Civil, Computer Science and Aeronautical Engineering
Roma Tre University
Rome, Italy

ISSN 0254-1971 ISSN 2309-3706 (electronic)
CISM International Centre for Mechanical Sciences
ISBN 978-3-031-84487-4 ISBN 978-3-031-84485-0 (eBook)
https://doi.org/10.1007/978-3-031-84485-0

This Springer imprint is published by the registered company Springer Nature Switzerland AG
The registered company address is: Gewerbestrasse 11, 6330 Cham, Switzerland

Preface

The tremendous impact of natural hazards, such as earthquakes, tsunamis, flooding, etc., which triggered technological accidents, referred to as natural-technological (NaTech) events, was demonstrated, for instance, by the recent Tohoku earthquake and the following Fukushima disaster in 2011 or by the UK's 2015 winter floods which topped £ 5bn, with thousands of families and businesses that faced financial problems because of inadequate or non-existent insurance. The Natech problem is quite relevant as up to 10% of industrial accidents, involving the release of Chemical, Biological, Radiological, Nuclear and high yield Explosives (CBRNE) substances, were triggered by natural hazards. To implement and support the Seveso III Directive 2012/18/EU, which regulates the control of major accident hazards involving dangerous substances, several aspects need to be analyzed. In this respect, this book aims to offer to students, scholars, and practitioners a clear overview of problems, available solutions and tools. The course, which was led by leading experts in Resilience and Natech, was a unique chance to learn about this hot topic developed by the resilience and risk community.

The book is subdivided in six different chapters. Chapter 1 by Ana Maria Cruz of Kyoto University (Japan) introduces the Natech risk problem relevant to both hazardous facilities and communities with a clear analysis of the current status and the future challenges. Chapter 2 by Roberto Nascimbene of IUSS Pavia (Italy) is focused on the seismic vulnerability of industrial plant equipment to seismic hazard and, in particular, of storage tanks and silos and the applicability of base isolation systems. Chapter 3 by Federica Ricci and Valerio Cozzani of the University of Bologna (Italy), is devoted to Natech risk assessment of process plants. More specifically, existing approaches are discussed, and the role of main ingredients, i.e. hazard, vulnerability and consequences, are identified. Critical aspects, e.g. cascading events, related to the consequences of damage caused by natural events, e.g. earthquakes, are also discussed and, examples of process and oil and gas industry are studied. Moreover, the role of safety barriers in major hazard industrial plants used for resilience enhancement is investigated.

Chapter 4 by Nicola Tondini, University of Trento (Italy), deals with Fire engineering and applications to industrial plant accidents. In particular, the formulation of the new concept of Performance-based fire engineering is introduced and the application to vulnerable industrial equipment like piperack devoted to support pipes is illustrated and discussed. In Chap. 5, by Fabrizio Paolacci et al., Roma Tre University (Italy), resilience concepts are formulated for industrial plants subjected to Natech events. A new general method for the quantification of resilience indeces like economic losses and business continuity is introduced and applied to process industry examples subjected to natural events. As illustrative example a Black Carbon plant subjected to seismic loading is analyzed and the results discussed. In Chap. 6 and last chapter by Oreste S. Bursi, University of Trento (Italy), a computational framework for an efficient and accurate seismic state-dependent fragility analysis is proposed; and it is applied to a full-scale prototype industrial steel frame structure. Also innovative metamaterial-based solutions for disaster reduction in facilities are proposed. To achieve this scope, the research was built upon the innovative scientific and technological concepts developed within the field of locally resonant metamaterials. Examples of the proposed concept are metafoundations, capable of rendering industrial storage tanks fully or partially isolated from the effects of incident waves striking from all directions.

Bologna, Italy — Valerio Cozzani
Rome, Italy — Fabrizio Paolacci

Contents

Chapter 1
Understanding Natech Events and Consequences at Hazardous Facilities

Ana Maria Cruz

1.1 Introduction

In today's interdependent and complex urban environments, it is not uncommon to have industrial facilities, fuel storage terminals, transportation infrastructure, water bodies and waterways, and so on, near urban centers and residential areas. This is the case in several megacities around the world. Figure 1.1, shows a photograph of a typical intersection in Tokyo. In the red circles, you may note the presence of a large fuel tank, and several fishing boats and a fuel tanker on the waterway. Tokyo is subject to typhoons and storm surges every year, and may be subject to large earthquakes and their possible tsunamis in the near future. Strong ground motion could result in liquid sloshing leading to an oil spill tsunami waves or strong storm surge could lead to the drifting of boats that could collide with other infrastructure causing damage and even resulting in oil spills. In 2018, Typhoon Jebi entered Osaka Bay causing extensive flooding at Kansai International Airport and the collision of a fuel tanker (which had lost its mooring due to the strong winds and storm surge) into the only bridge that connects the airport to Osaka City. While there was no oil spill in this case, as the tanker had already unloaded the fuel at the airport, the accident provided a clear example of potential accident chains that could lead to natural hazard-triggered technological accidents with the consequent loss of containment of hazardous materials (known as Natechs).

The aim of this chapter is to provide the reader with an understanding of Natech events and their potential consequences at hazardous facilities. Furthermore, based on evidence from past Natech events and research studies, the chapter argues that there is a need for a paradigm shift in Natech risk management, and introduces a comprehensive, area-wide risk management framework.

A. M. Cruz (✉)
Disaster Prevention Research Institute, Kyoto University, Kyoto, Japan
e-mail: cruznaranjo.anamaria.77m@st.kyoto-u.ac.jp

V. Cozzani and F. Paolacci (eds.), *Management Strategies and Resilience Against NaTech Events*, CISM International Centre for Mechanical Sciences 616,
https://doi.org/10.1007/978-3-031-84485-0_1

Fig. 1.1 A busy intersection in Tokyo showing transportation infrastructure over water, air and land. Highlighted in the red circles are a fuel storage tank in the background, and fishing boats on the canal in the foreground

The rest of the chapter is organized as follows. Section 1.2 discusses the question of whether Natech risk is really a problem. Here, the incidence of Natechs events is also briefly discussed, as well as the evidence so far concerning the influence of climate change on Natech incidence. Section 1.3 presents several examples from past Natech events and discusses problems in their risk management. Section 1.4 discusses the need for a paradigm shift in Natech risk management and proposes a comprehensive, area-wide framework. Section 1.5 presents some conclusions and future research directions.

1.2 Is There a Problem?

So, is there a problem? Several studies indicate that there could be a problem for various reasons, including increased population in highly urbanized areas subject to natural hazards, and industrialization. A recent study published in Nature found that the world population exposed to tropical cyclones increased from 408 million to 792 million people annually over the period 2002 to 2019 (Jing et al. 2024). According to Jing et al. (2024), this increase in global exposure to tropical cyclones is primarily driven by Asia, although increases were noted in Europe and elsewhere as well. Currently, an overall global increase (Statista 2024) in chemicals production is forecasted in Asia, with the highest growth expected in China (BASF 2023). At the same time, the number of natural and technological disasters around the globe have increased, with Asian countries, in particular, registering the highest number of natural disasters.

This increasing trend is also true for Natech events. In the work by Luo et al. (2020), Natechs represented 3–7% of all accidents reported in the United States between 1990 and 2017. After accounting for changes in regulations and reporting requirements, and other factors, an increase in the number of Natechs events over the same period was confirmed. Furthermore, spatial temporal analysis of hydrometeorological related Natechs events indicated that climate change may be an important factor in the increasing trend (Luo et al. 2021).

Climate change is indeed an important factor, not only because it is affecting weather patterns around the world, but also because climate change may exacerbate pre-disaster conditions. This was the case when a Mw 6.6 earthquake struck Hokkaido, Japan, on September 6, 2018, just a day and a half after the passing of Typhoon Jebi through the area. There were many earthquake-triggered landslides, probably the result of colluvial slope failures. These slopes were most likely preconditioned for failure due to the heavy precipitation from the typhoon (Wang and Nam 2023).

1.3 Learning from Past Accidents

1.3.1 Earthquakes

Strong ground motion caused by earthquakes poses threats to buildings and structures at industrial installations. The vulnerability of buildings and structures will depend on the hazard, the type of equipment, materials, operation conditions, and most importantly on the adoption of appropriate building codes (or their retrofitting if structures are old) (Wisner et al. 2012). The following examples point to challenges in risk management when addressing Natech events related with earthquakes.

The Kocaeli Earthquake in Turkey in 1999. The Kocaeli earthquake of 1999 was one of the first major earthquakes, Mw7.4, to affect a highly industrial area (Steinberg and Cruz 2004). The earthquake killed thousands of people and left hundreds of thousands homeless. Over 21 hazardous materials releases with offsite consequences were documented by Steinberg and Cruz (2004). One of the major accidents involved the release of acrylonitrile caused by liquid sloshing and elephant foot buckling of storage tanks at an acrylic fiber plant in Yalova. The simultaneous loss of electricity, damage to containment dikes, and damage to fire water and foam systems contributed to slow emergency response and containment (Cruz and Steinberg 2005). The authors also noted that the accident resulted in water, ground water, soil and air contamination and pollution. The toxic plume from the release resulted in the evacuation of residents living near the installation at a time when search and rescue in the area were still ongoing.

A survey of affected installations following the Kocaeli earthquake (Cruz and Steinberg 2005) showed that chemical releases occurred not only due to structural damage (some installations suffered complete collapse due to lack of earthquake resistant structures) but also due to non-structural failures (Cruz and Steinberg 2005). The concurrent damage to and loss of electrical power systems and other critical infrastructure actually led to releases or hampered emergency response and containment. This was the case at a fertilizer plant in Korfez. Loss of electricity and insufficient back up power resulted in loss of cooling, leading to over-heating, and the intentional release of ammonia anhydrous into the atmosphere to avoid overpressurization and a possible explosion. The authors also found that a significantly higher number of industrial facilities surveyed, built prior to the introduction of the first seismic building codes in Turkey in 1975, suffered damage. This disaster highlighted the area-wide impacts of the earthquake which led to multiple chemical accidents. It also highlighted the importance of building industrial infrastructure to the latest seismic design codes, but also the need to consider natural hazard effects on safety barriers and systems, mitigation measures, and emergency response.

The Wenchuan Earthquake in China in 2008. In 2008, the Wenchuan earthquake of magnitude Mw 7.8 caused extensive damage killing over 70,000 people. The earthquake caused heavy damage to industrial installations, particularly older building structures that collapsed killing workers, and damaging industrial equipment leading to chemical spills (Krausmann et al. 2010b). Krausmann et al. (2010b) found that the damage severity of installations visited correlated well with the age of the plant (R: 0.79, p value: 0.0007), with older facilities suffering more damage than more recent ones having adopted newer building design codes (Krausmann et al. 2010b). Similar to the previous case, the earthquake triggered chemical accidents at multiple installations, while also hampering the ability of responders to contain them.

The Great East Japan Earthquake. In 2011, the Mw 9.0 earthquake and tsunami that struck the Pacific east coast of Japan resulted in more than 1400 reported chemical releases (Nishi 2011), with a few of them severe (Krausmann and Cruz 2013b; Chakraborty et al. 2018). The most notable examples include:

- Earthquake triggered damage caused fire and explosions at a liquified petroleum gas (LPG) storage tank farm at a refinery in Ichihara City, Chiba Prefecture which took more than 10 d to extinguish, and required evacuation of residents in an area about 2 km in radius from the accident site.
- Oil spills, fire, and floating fires at a refinery in Sendai due to a chain of events induced by the tsunami which took several days to contain and resulted in the evacuation orders even for those already at tsunami evacuation shelters (Yu et al. 2017).
- A gasoline spill at a fuel storage terminal in Sendai caused by floating debris impact on a purge line. The release was only detected on the third day after the earthquake, and after the contents of the large storage tank had been released,

posing a fire threat to a neighboring LNG regasification plant and storage tank supplying gas to Sendai city. The event resulted in yet another evacuation order (Chakraborty et al. 2018; Yu et al. 2017). In fact, according to Yu et al. (2017), some residents evacuated up to 4 times due to the tsunami and Natech accidents in Sendai.

- The tsunami caused the floating off and spill of oil storage tanks in the Bay of Kesennuma resulting in a large floating fire that was spread by the tsunami waves (Nishino and Takagi 2020)

The Great East Japan earthquake was a multi-hazard event, triggering a huge tsunami, liquefaction, land subsidence, landslides, chemical accidents including fires and explosions, and the Fukushima nuclear accident. Following the disaster, the need to prepare for expected events (within the design level of infrastructure systems), but also for events that may exceed design levels has been discussed. A two-tier system has been introduced when planning for future tsunami disasters (Aoki 2021). In the case of major events that exceed design levels (e.g., events with return periods of 1000 years), disaster education, land use planning, disaster preparedness, early warning systems, and evacuation should be prioritized.

New Year's Earthquake in Noto Peninsula in 2024. The Mw7.6 earthquake that struck the Noto Peninsula in Japan on 1 January 2024 caused extensive damage to residential houses, manufacturing industry, and transportation infrastructure. Two nuclear power plants in the region were exposed to strong ground motions. Both facilities reported that there were no abnormalities, except for damage to a transformer at the Shika Nuclear Power plant where the highest ground shaking for the area was recorded (Japan Times 2024). Following the earthquake, local authorities in Shika town have requested that the government re-evaluate the emergency response guidelines for Nuclear accidents. Had there been a nuclear accident at the plant, it would have been difficult to inform residents living in potentially exposed areas due to loss of electricity and loss of communication systems including no phone or mobile phone services, nor internet services. Furthermore, almost all the roads in the area were severely damaged leaving communities isolated. Discussions are ongoing to revise the guidelines (Japan Times 2024).

1.3.2 Tropical Cyclones

Tropical cyclones including hurricanes and typhoons can cause extensive damage to buildings and structures, including those at industrial installations (Cruz and Steinberg 2005). Tropical cyclones are multi-hazard events which may include high winds, tornadoes, heavy rain, flooding, storm surge and coastal flooding, and lightning. There are several examples of tropical cyclone related Natech events.

Hurricane Katrina and Rita. In 2005, Hurricane Katrina and Rita, caused hundreds of oil and gas spills from offshore and fixed facilities in the US Gulf of Mexico (Cruz and Krausmann 2009). In one case, a large oil storage tank, probably damaged by floating debris, caused the spill of about 8000 barrels of oil, affecting more than 70,000 homes in Chalmette, Louisiana.

Hurricane Harvey. In 2017, following the passing of Hurricane Harvey, a fire broke out at the Arkema Plant in Crosby, Texas. According to the U.S. Chemicals Safety Board (2018), the fire broke out due to over-heating and spontaneous ignition of organic peroxide-containing storage tanks that had lost their cooling due to power outage caused by the storm. The accident required the evacuation of residents living within a 2.4 km radius from the plant.

Typhoon Jebi. In 2018, Typhoon Jebi passed over Osaka Bay, causing the loss of mooring and collision of several ships, including an oil tanker that collided with the only bridge connecting Osaka City to the Kansai International Airport, which was also flooded (Mainichi 2018). The high storm surge also affected the port area in Nishinomiya City, causing the drifting of hundreds of vehicles and containers and several fires (BBC 2018).

Typhoon Faxai. In 2019, Typhoon Faxai made landfall in Japan, passing over the industrial parks in Chiba Prefecture in the Tokyo Bay area. The typhoon's high winds caused several problems including (Cruz et al. 2020):

- The collapse of a cooling tower at the Japan Atomic Energy Agency Oarai in Ibaraki Prefecture. According to a report (Japan Times 2024), the cooling tower collapsed due to deterioration and rotting of the internal wood structure of the cooling tower, lack of adequate inspection and structural integrity check, and the strong wind.
- A stripping tower of a gas converter collapsed at a steel mill at an oil complex in Kimitsu City. The collapse of the stripping tower resulted in damage to the dust-collecting pipes and gas pipes of the converter. However, no gas leak and no injuries were reported. According to local authorities, there was no risk of fire.
- A pipe below a storage tank containing hydrochloric acid (concentration: 17%) was damaged by a steel step ladder that fell due to high winds during the typhoon at a chemical company. The damage caused leakage of hydrochloric acid. The released hydrochloric acid mixed with rainwater and flowed out from a drain to the sea. The concentration of the hydrochloric acid and rain water solution was measured and confirmed to be $<1\%$ in the drain.
- A steel ladder on a flare stack fell due to the strong wind at an industrial installation in Ichihara City. There was no gas release or major damage due to the ladder collapse. Damage to scaffoldings and insulation in several areas of the plant was also reported. Furthermore, a small truck was lifted and turned over by the wind, and a temporary warehouse at a plant in the Sodegaura District was damaged. No injuries were reported.

The installations above, affected by Typhoon Faxia, were within approximately 10 km from the center of the typhoon's path, on the front right quadrant, which experiences the highest wind speeds. An anemometer located near the Chiba Prefecture

Police Headquarters measured a maximum wind speed of 53 m/s, while maximum wind speeds of 42.6 m/s were measured at industrial parks at Anegasaki District and Sodegaura District where most damage occurred (Cruz et al. 2020). According to company officials in the area, the plants in both districts follow the Building Standards Law in Japan (design base: 38 m/s–10 min average wind speed at 10 m height) based on the Provisions of the Revised Building Code). Wind speeds associated with Typhoon Faxai, which had been downgraded from a Super Typhoon to a Typhoon when passing through the area, still exceeded design levels. This prompted concern from company officials, particularly when given the possibility of stronger typhoons making landfall in the Tokyo Bay area due to climate change.

1.3.3 Flooding

Flooding of industrial installations can cause floating off, displacement and overturning of storage tanks and pipelines, water intrusion, damage to low-lying equipment, and damage to storage tanks and other equipment due to the impact of floating debris, which can lead to hazardous materials releases. As described earlier, tropical cyclones can cause heavy rain and storm surge resulting in flood-related Natech accidents. In recent years, floods have also resulted in Natech accidents from non-chemical process industries in Japan, but that pose tremendous threats to nearby communities. Some examples are described below.

West Japan Floods (2018). A vapor explosion occurred at an aluminum recycling plant in Shimobara town, Soya City, Okayama Prefecture, in 2018, when flood waters entered in contact with molten aluminum in the furnace (Araki et al. 2021). Fortunately, there were no fatalities and only some minor injuries. Nevertheless, several problems were noted. For example, workers at the plant did not inform residents nor first responders of the threat of an explosion despite the fact that they were well aware of the danger and evacuated the plant about an hour before the explosion, when water reached their knees. The explosion occurred at around 11:30 p.m. Neither residents nor Soya City fire department had any knowledge that there could be a vapor explosion at the aluminum plant. A prompt evacuation was improvised thanks to the fact that the community had carried out disaster drills regularly and were well organized and that Soya City and Soya fire department helped to coordinate the efforts.

Floods in Saga Prefecture (2019). An oil spill at an ironworks company in Omachi town, Saga Prefecture in 2019, occurred when flood waters entered the plant (Misuri et al. 2021a). Similar to the situation in Shimobara town, there was little to no disaster preparedness planning that considered potential for oil spills, despite the fact that a similar event had occurred at the same facility several years before. Interviews with residents showed that some residents evacuated vertically because they felt that similar to previous occasions, flood waters would recede quickly or would be pumped to the Rokkaku river promptly. However, pumping the flood water to the river was not possible as it would contaminate the river (Misuri et al. 2021a).

Floods in Kumamoto Prefecture (2020). A phreatic explosion and fire occurred at a graphitization furnace at a carbon manufacturing plant in Ashikita district, Kumamoto Prefecture in 2020, when flood waters came up from under the ground and entered into contact with a furnace . Evacuation orders were issued, however, the area was already flooded. This was a problem for people who had decided for vertical evacuation prior to the floods.

It is important to note that the above facilities fall outside of stricter regulations for the handling of hazardous materials in Japan. Therefore, there was little to no disaster preparedness planning for these types of events among the local government authorities and the fire departments. Furthermore, there was a lack of information regarding potential accidents and their consequences disclosed to the public (Araki et al. 2021; Misuri et al. 2021a).

Overall, the above examples highlight many of the challenges posed by Natech accident events. Area-wide impacts and multiple accidents are possible when natural hazard events affect large areas. Furthermore, accidents are not always initiated due to damage within the fence line of facilities, but due to impacts or damage to other infrastructure. Accidents can also occur, and quite often, due to non-structural failures or failure of critical infrastructure. Furthermore, natural hazard events affect safety barriers, and mitigation and emergency response measures. Finally, natural hazards are often "multi-hazard," making the risk assessment of Natech events more complex. Due to the area-wide and multi-hazard impacts of natural hazards, Natech risk management requires a new approach to risk assessment and risk management.

1.4 The Need for a Paradigm Shift

Natech risk reduction requires "a paradigm shift in risk management", where the focus of analysis of these hazards moves beyond the individual industrial facility's fence line to an area-wide or territory-wide scope." That is the "system" is no longer only one industrial facility, but that the territory, including the facility, neighboring industries and businesses, interconnected infrastructure and surrounding residential areas, are, in fact, the system! Furthermore, the need for flexibility in order to adapt to rapidly changing dynamic situations is imperative.

Suarez-Paba and Cruz (2022) proposed a comprehensive, area-wide framework that seeks to enhance industries' resilience to Natech events by considering the wider-social and environmental context in which industrial installations are located. The proposed framework is divided into four categories including infrastructure, organizational and safety management, external environment, and risk governance and risk communication. Risk communication is key, and it should happen throughout the whole process. The biggest difference in the proposed framework from traditional risk management frameworks, is the need to involve stakeholders from outside individual organizations. Through multi-stakeholder discussion and participation, the potential interactions between the natural and built environment in a territory can be better understood and assessed. Multi-stakeholder dialogue needs to be supported through risk governance mechanisms, particularly in areas with mixed land uses, and the presence of both private and public land and infrastructure. Pursuing participation from all stakeholders not only contributes to the buy-in of agreed-upon risk reduction strategies and cooperation but also improves overall preparedness against such disasters.

1.5 Conclusions

This chapter has presented examples of Natech events with the aim of improving the reader's understanding of Natech events and their consequences, and the challenges for risk management.

Risk management of Natech hazards requires a paradigm shift taking into consideration the following points:

- Natural hazards may affect large areas, thus, an area-wide scope beyond individual facilities is needed for comprehensive Natech risk management
- Natech events are likely to occur when natural hazards exceed design levels. Thus, there is the need to incorporate notions of resilience engineering and systems to ensure territorial resilience and flexibility in the face of rapid environmental changes.
- Multi-stakeholder dialogue for risk reduction by local governments, companies and residents is key.
- The disclosure of hazard and/ or risk information is crucial for disaster preparedness in order to move toward territorial resilience and risk reduction.

Finally, it is necessary to propose and implement precautionary measures and enhance risk awareness to cooperatively build on strategies that help diminish Natech impacts and advocate for more resilient societies.

References

Aoki, S. (2021). Natech risks in Osaka bay area and collaborative activities for risk reduction. In *Oral presentation, 5th Natech Symposium*, Osaka, Japan, 10–11 March 2021.

Araki, Y., Hokugo, A., Pinheiro, A. K., Ohtsu, N., Cruz, A. M. (2021). Explosion at an aluminum factory caused by the July 2018 Japan floods: Investigation of damages and evacuation activities. *Journal of Loss Prevention in the Process Industries, 69*, 104352.

BASF. (2023). Outlook for the chemical industry. BASF report 2023. Shareholders. Technical report, 2023. Retrieved February 23, 2024.

BBC. (2018). *Japan's Typhoon Jebi leaves destruction in its wake*. https://www.bbc.com/news/world-asia-45419771

Chakraborty, A., Ibrahim, A., Cruz, A. M. (2018). A study of accident investigation methodologies applied to the Natech events during the 2011, Great East Japan earthquake. *Journal of Loss Prevention in the Process Industries,51*(208–222).

U.S. Chemicals Safety Board. *Organic peroxide decomposition, release, and fire at Arkema crosby following hurricane harvey flooding*. Report Number: 2017-08-I-TX. 2018. ///Users/anamariacruz/Downloads/*FINAL_Arkema_draft_report*_2018-05-23.pdf.

Cruz, A. M., & Krausmann, E. (2009). Hazardous-materials releases from offshore oil and gas facilities and emergency response following hurricanes Katrina and Rita. *Journal of Loss Prevention in the Process Industries, 22*(1), 59–65

Cruz, A. M., Ishimaru, H., & Aoki S. (2020). Damage and chemical releases from industrial facilities affected by Typhoon 15 in 2019. Section 3-2-5 in Research report: Comprehensive survey on the effects of prolonged power outages and wind and flood damage caused by Typhoon No. 15 in 2019. (In Japanese: 令和元年 台風15号による停電の長期化に伴う影響と風水害に関する総合調査).

Cruz, A. M., Ishimaru, H., & Damage, A. S. (2020). Damage and chemical releases from industrial facilities affected by typhoon 15 in 2019. section 3-2-5 in research report: Comprehensive survey on the effects of prolonged power outages and wind and flood damage caused by typhoon no. 15 in 2019.

Japan Times. (2024). *Noto quake exceeded projections of nuclear plant operator*. https://www.japantimes.co.jp/news/2024/02/03/japan/shika-nuclear-plant-damage/

Jing, R., Heft-Neal, S., Chavas, D. R., Griswold, M., Wang, Z., Clark-Ginsberg, A., Guha-Sapir, D., Bendavid, E., Wagner, Z. (2024). Global population profile of tropical cyclone exposure from 2002 to 2019. *Nature, 626*(7999), 549–554.

Krausmann, E., & Cruz, A. M. (2013). Impact of the 11 March 2011, Great East Japan earthquake and Tsunami on the chemical industry. *Natural Hazards,67*, 811–828.

Krausmann, E., Cruz, A. M., & Affeltranger, B. (2010b). The impact of the 12 may 2008 Wenchuan earthquake on industrial facilities. *Journal of Loss Prevention in the Process Industries, 23*(2), 242–248.

Luo, X., Cruz, A. M., & Tzioutzios, D. (2020). Extracting Natech reports from large databases: development of a semi-intelligent Natech identification framework. *International Journal of Disaster Risk Science, 11*(6), 735–750.

Luo, X., Cruz, A. M., Tzioutzios, D. (2021). Climate change and temporal-spatial variation of tropical storm-related Natechs in the united states from 1990 to 2017: Is there a link? *International Journal of Disaster Risk Reduction,62*(102366).

Misuri, A., Cruz, A. M., Park, H., Garnier, E., Ohtsu, N., Hokugo, A., Fujita, I., Aoki, S.-I., & Cozzani, V. (2021a). Technological accidents caused by floods: The case of the saga prefecture oil spill, Japan 2019. *International Journal of Disaster Risk Reduction, 66*, 102634.

Nishi, H. (2011). Damage on hazardous materials facilities. In *Proceedings of the International Symposium on Engineering Lessons Learned from the 2011*, Great East Japan Earthquake, March 1–4, 2012, Tokyo, Japan.

Nishino, T., & Takagi, Y. (2020). Numerical analysis of tsunami-triggered oil spill fires from petrochemical industrial complexes in Osaka Bay, Japan, for thermal radiation hazard assessment. *Int. J. Disaster Risk Reduction, 42*, 101352.

Statista. (2024). *Market Insights Industry Manufacturing Material Products Chemicals*. https://www.statista.com/outlook/io/manufacturing/material-products/chemicals/asean

Steinberg, L. J., & Cruz, A. M. (2004). When natural and technological disasters collide: Lessons from the Turkey earthquake of August 17, 1999. *Natural Hazards Review, 5*(3), 121–130.

Suarez-Paba, M. C., & Cruz, A. M. (2022). A paradigm shift in Natech risk management: Development of a rating system framework for evaluating the performance of industry. *Journal of Loss Prevention in the Process Industries,74*, 104615.

The Mainichi. (2018). *Flooded Kansai airport had crucial infrastructure on basement floor*. https://mainichi.jp/english/articles/20180911/p2a/00m/0na/032000c

Wang, F., & Nam, K. (2023). Landslide disasters caused by the 2018 Eastern Iburi earthquake in Hokkaido Japan and the countermeasures to completely prevent the similar disasters in the future. In *Progress in landslide research and technology* (Vol. 1 Issue 1, pp. 305–313). Springer.

Wisner, B., Gaillard, J.-C., & Kelman, I. (2012). *Handbook of hazards and disaster risk reduction*. Routledge.

Yu, J., Cruz, A. M., & Hokugo, A. (2017). Households' risk perception and behavioral responses to Natech accidents. *International Journal of Disaster Risk Science, 8*, 1–15.

Chapter 2
Vulnerability of Industrial Plant Equipment to Natural Hazards

Roberto Nascimbene

2.1 Introduction

The vulnerability of industrial plant equipment, specifically tanks and silos, to natural hazards poses a significant challenge, particularly in regions prone to earthquakes. The potential for damage during seismic events stems from the dynamic forces and ground shaking that these structures must endure. The seismic vulnerability of tanks and silos is contingent upon various factors including the structural design, soil conditions, and the seismicity of the area.

Ensuring the resilience of these structures involves strict adherence to seismic design codes and construction standards. Engineers must carefully consider factors such as material selection, reinforcing structures, and the adequacy of foundation design. Foundation integrity plays a pivotal role in the seismic resistance of tanks and silos, necessitating designs that account for local soil conditions and provide robust support.

For older industrial facilities, retrofitting and upgrading become imperative to meet current seismic standards and enhance the overall resilience of tanks and silos. Incorporating flexible connections between different structural components allows for some movement during seismic events without compromising stability. Emergency planning is essential, encompassing procedures for operational shutdown, personnel evacuation, and the assessment of potential structural damage in the aftermath of an earthquake.

Conducting thorough risk assessments helps identify vulnerabilities and potential failure modes, aiding in the prioritization of mitigation measures and resource

R. Nascimbene (✉)
University School for Advanced Studies IUSS Pavia, Pavia, Italy
e-mail: roberto.nascimbene@iusspavia.it

V. Cozzani and F. Paolacci (eds.), *Management Strategies and Resilience Against NaTech Events*, CISM International Centre for Mechanical Sciences 616,
https://doi.org/10.1007/978-3-031-84485-0_2

allocation. Implementing monitoring systems that provide real-time data on the condition of tanks and silos is crucial for early detection of issues and timely preventive measures. Regulatory compliance with local and national seismic design standards is paramount, requiring industries to stay abreast of any updates or changes in regulations.

Collaboration with seismic engineering experts is highly beneficial, as their expertise can offer valuable insights into the specific risks and vulnerabilities associated with industrial equipment in earthquake-prone areas. This collaborative approach, encompassing proper design, construction practices, monitoring, and emergency planning, is essential to mitigate the vulnerability of industrial plant equipment to natural hazards, especially in regions with a high seismic risk.

In the subsequent sections, we will delve into key aspects that contribute to the seismic resilience of industrial plant equipment, specifically focusing on tanks and silos: Sect. 2.2 Codes and guidelines; Sect. 2.3 Damage to tanks induced by seismic events; Sect. 2.4 Vulnerability of cylindrical elements; Sect. 2.5 Effects of soil-structure interaction; Sect. 2.6 Base-isolation devices. In conclusion, the examination of codes and guidelines, damage mechanisms, vulnerability of cylindrical elements, soil-structure interaction, and the potential benefits of base-isolation devices will provide a comprehensive understanding of the seismic considerations crucial for the robust design and retrofitting of industrial plant equipment in earthquake-prone regions.

2.2 Codes and Guidelines

Ensuring the seismic integrity of industrial equipment necessitates strict adherence to established codes and guidelines. These regulatory frameworks outline the minimum standards for seismic design and construction, encompassing aspects such as material specifications, structural configurations, and foundation requirements. A comprehensive understanding of and compliance with these codes are fundamental to mitigating seismic vulnerabilities effectively.

The following is an overview of various international codes, standards, guidelines, and recommendations pertinent to the seismic design of industrial tanks and silos, highlighting key details from specified documents:

- **UNI EN 1998-4:2006 (EC8-4)**:
  - **Focus**: Design of structures for earthquake resistance, specifically silos, tanks, and pipelines.
  - **Scope**: Provides principles, analyses, and design rules for seismic evaluation of industrial facilities, covering above ground and buried pipelines, storage tanks, and silos of various types and uses.
  - **Inclusion**: Criteria and rules encompass sizes, structural materials (reinforced concrete, prestressed concrete, steel), types, stored materials (fluid or granular), and soil-structure interaction.

  - **Reference**: UNI EN 14015:2006 for the seismic design of storage tanks based on API 650.

- **Seismic Design of Storage Tanks (NZSEE-09)**:

  - **Origin**: New Zealand National Society for Earthquake Engineering.
  - **Revision**: Aligns with New Zealand loading code NZS 1170.5 (2004) "Structural design actions."
  - **Consideration**: Examines tanks as described in UNI EN 1998-4:2006, providing recommendations for seismic design.

- **ACI 350.3-06**:

  - **Issuer**: American Concrete Institute (ACI).
  - **Coverage**: Seismic design of liquid-containing concrete structures, including rectangular and circular tanks (on-grade and below grade).
  - **Additional**: ACI 371R-98 guides the analysis, design, and construction of concrete-pedestal water towers.

- **API 650 2012 Edition**:

  - **Issuer**: American Petroleum Institute (API).
  - **Purpose**: Establishes requirements for material, design, fabrication, erection, and testing of welded storage tanks for petroleum products.
  - **Seismic Consideration**: Appendix E provides seismic design requirements for welded on-grade steel flat-bottom tanks.

- **API 620 2012 Edition**:

  - **Issuer**: American Petroleum Institute (API).
  - **Distinction**: Covers higher pressure and temperature than API 650 (atmospheric storage tanks).
  - **Seismic Aspect**: Appendix L outlines minimum requirements for the seismic design of welded storage tanks.

- **AWWA D100 2011 Edition and AWWA D103 Edition 2009**:

  - **Issuer**: American Water Works Association (AWWA).
  - **Focus**: Minimum requirements for design, construction, inspection, and testing of welded and factory-coated bolted steel tanks for water storage.
  - **Related**: AWWA D115 2006 Edition provides recommendations for tendon-prestressed concrete water tanks.

- **ASCE 7 2010 Edition**:

  - **Issuer**: American Society of Civil Engineers (ASCE).
  - **Inclusion**: Nonbuilding structures resisting earthquake effects with Chap. 15 emphasizing tanks and vessels of various materials.

- **IITK-GSDMA (2007)**:
  - **Source**: Indian Institute of Technology Kanpur (IIT Kanpur)—Guidelines for Seismic Design of Liquid Storage Tanks.
  - **Content**: Provides provisions, commentary, and explanatory examples. Reviews and compares several international codes for seismic analysis of tanks.
- **COVENIN 3623 2000 Edition**:
  - **Origin**: National Code from Venezuela.
  - **Purpose**: Establishes minimum requirements for seismic-resistant design and analysis of vertical and cylindrical steel storage tanks.
- **NCh2369 2003 Edition**:
  - **Origin**: National Code from Chile.
  - **Emphasis**: Chap. 11.7 and 11.8 focus on above ground and elevated tanks, referencing American standards.
- **AIJ Design Recommendation 2010 Edition**:
  - **Issuer**: Architectural Institute of Japan (AIJ).
  - **Application**: Structural design, especially seismic design of water storage tanks, silos, and other storage structures.

While these codes and standards address seismic analyses of liquid-containing tanks, their implementation strategies vary, leading to significant differences in design forces. A comprehensive review and comparison have been conducted to highlight similarities, discrepancies, and limitations across these documents. Jaiswal et al. (2007) have extensively reviewed and compared ten specific documents to elucidate these variations in design, analysis, and modeling of tanks.

2.3 Damage to Tanks Induced by Seismic Events

An in-depth exploration of the potential damage inflicted upon tanks during seismic events is paramount. This includes the assessment of structural deformations, integrity breaches, and the likelihood of catastrophic failure. Understanding the specific modes of damage is essential for devising mitigation strategies and retrofitting measures that address vulnerabilities exposed during seismic loading.

The potential damage and failure of storage tanks during earthquakes can have far-reaching consequences, surpassing the costs associated with structural repair and reconstruction. The repercussions may extend to disruptions in public water supply, affecting both domestic consumption and post-earthquake recovery efforts. Loss of a tank can compromise fire protection precisely when it is most crucial. Moreover, tanks may contain hazardous materials, posing risks of environmental contamination or triggering immediate hazards like vapor and toxic gas clouds. Various failure modes induced by ground motion can manifest diversely and they are commonly categorized as follows:

1. Elastoplastic Buckling ("Elephant's Foot" Buckling—Fig. 2.1a): Common in steel tanks; outward bulging above the base forming an "elephant's foot." Results from circumferential tensile stresses induced by internal hydrostatic and hydrodynamic pressures coupled with axial compression from seismic loads. Typically observed in large-size tanks with a low height-to-radius ratio.
2. Elastic Buckling ("Diamond-Shaped" Buckling—Fig. 2.1b): A form of elastic instability near the base of tanks with higher vertical compression. Occurs in more slender tanks with a height-to-radius ratio greater than one. Elastic buckling considerations may govern design for cylindrical shells with low internal pressure. During the Emilia Earthquake Sequence (2012) uncommon diamond-shape elastic buckling observed near circumferential welding related to imperfections induced by welding and variations in shell thickness (Fig. 2.1c).
3. Local Buckling in Upper Areas: Local buckling observed in upper areas of the shell, possibly due to rupture near the base followed by rapid loss of tank contents. Associated with reduced thickness near the top of the tank shell.
4. Relative Displacements and Connection Failures (Fig. 2.1d): Displacements between tank and connected piping/equipment induce damage to connections. While not a direct tank damage, it causes operational breaks and economic losses.
5. Sloshing Waves and Roof Failure: Insufficient freeboard can lead to upward pressure on the tank roof. May cause roof failure or damage to the shell-roof joint due to sloshing waves impacting the roof.
6. Explosion Risk in Inflammable Material Tanks: Bouncing of floating roof against the shell can produce sparks, potentially igniting inflammable material.
7. Liquefaction and Foundation Settlement: Tanks in areas prone to liquefaction may experience global or partial settlement of the foundation. Sliding or walk-off of foundations due to progressive rocking is observed.
8. Damage to Connection Between Tank and Foundation (Fig. 2.1e): Fracture or extraction of anchor bolts, spalling of concrete, and flexural failure in anchor plates. These damages are not related to the shell structure but result from tank-induced actions and poor foundation design.
9. Issues in Leg-Supported Tanks (Fig. 2.1f): Problems in legs or their connections, especially in cases of a "strong leg-weak tank wall" mechanism. Unanchored legged systems often perform better, exhibiting a form of not-designed base-isolation response.

Understanding these diverse failure modes is crucial for implementing effective seismic design and retrofitting measures to enhance the resilience of storage tanks and mitigate potential hazards and economic losses.

The imperative to fortify the seismic resilience of storage tanks arises from a multifaceted set of considerations, underscoring the overarching objectives of safety, environmental protection, and economic stability. One of the foremost drivers behind seismic improvements is the paramount goal of safeguarding public safety. The

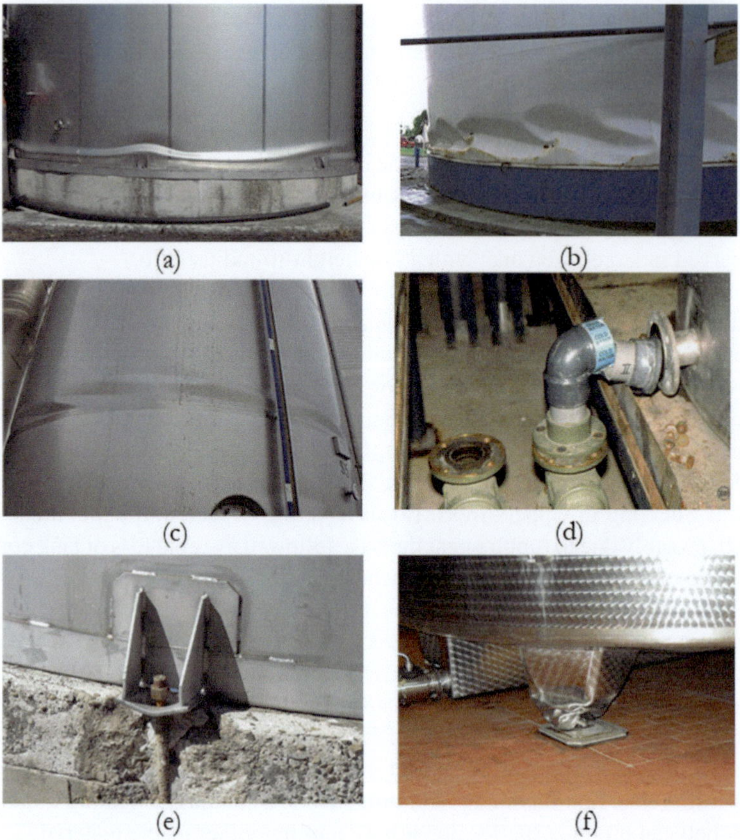

Fig. 2.1 Observed damages of tanks in past earthquakes: **a**, **c**, **e** and **f** Emilia earthquake Italy (May 20th and 29th 2012 Magnitude 6.1 and 5.9 respectively); **b** Izmit earthquake Turkey (August 17th 1999 Magnitude 7.4) (from Izmit Collection University of California Berkeley); **f** Edgecumbe earthquake New Zealand (March 2nd 1987 Magnitude 6.3) (from Edgecumbe Collection University of California Berkeley)

prevention of catastrophic failures during earthquakes is pivotal in averting potential hazards such as toxic releases, fires, or explosions, thereby ensuring the well-being of communities in proximity to these structures. Storage tanks commonly house materials that possess inherent environmental risks. The strategic improvement of seismic performance serves as a proactive measure to mitigate the possibility of spills, leaks, or releases. This not only prevents environmental contamination but also aligns with broader goals of environmental protection. As integral components of critical infrastructure, storage tanks play a pivotal role in providing essential resources such as water or fuel. Strengthening the structural integrity of tanks ensures operational continuity during and after seismic events. This, in turn, supports recovery efforts and maintains the uninterrupted provision of crucial services. Beyond safety and environmental concerns, there exists a significant economic dimension. Failures of storage tanks during earthquakes can result in substantial repair or reconstruction costs. Implementing improvement strategies becomes instrumental in minimizing these economic losses and expediting the recovery process.

In the subsequent section, we will delve into the proactive identification and remediation of vulnerabilities in tank structures. This strategic approach plays a crucial role in diminishing the overall seismic risk, thereby reducing the likelihood of failures and mitigating the subsequent consequences linked to such occurrences.

2.4 Vulnerability of Cylindrical Containments

Cylindrical elements, a prevalent structural feature in tanks and silos, warrant specific consideration due to their susceptibility to seismic forces. The analysis will encompass the examination of stresses, deformations, and failure modes in cylindrical structures subjected to seismic loading. Addressing the vulnerability of these elements is critical for enhancing the overall seismic resilience of industrial equipment.

The analysis of dynamic fluid pressures generated during an earthquake is crucial for the design of containment structures like tanks and silos. Evaluating a tank-fluid system necessitates considering the movements of both the structure and the enclosed liquid. In Fig. 2.2a, when a tank holding liquid undergoes lateral acceleration $a(t)$ at its base, the hydrodynamic pressures exerted by the fluid on the wall and base change in intensity and distribution from those observed under a state of hydrostatic equilibrium (Haroun 1980). For an upright containment structure excited horizontally, a portion of the fluid near the upper free surface tends not to displace laterally with the tank walls. Instead, it undergoes vertical sloshing or rocking oscillations about a horizontal axis normal to the direction of excitation (essentially antisymmetric modes) (Calvi and Nascimbene 2011). Closer to the floor base and along the walls at the bottom, the fluid is unable to move out of the way as the tank displaces, moving synchronously with the wall and acting as an added mass rigidly attached to the wall (Calvi and Nascimbene 2023). The fluid's motion in a storage container can be considered composed of two main parts: the first part that moves in unison with the tank, known as the rigid impulsive mass component, and the second part undergoing sloshing motion, known as the convective mass component. The proportions of the liquid mass participating in these two types of motion depend on the tank's geometry: the height of the free surface of the fluid (H), diameter (D = 2R), and radius (R). The portion of the fluid that acts in a convective fashion decreases as the tank aspect ratio ($\gamma = H/R$) increases, with the rigid mode becoming more predominant. In the case of tall/slender tanks, the pressure induced by the impulsive mass component is maximum near the tank base and is associated with high-frequency oscillations. Conversely, for tanks with very low aspect ratios, such as shallow/squat/broad tanks, only about 30% of the total fluid acts with the walls, with the remainder responding in sloshing modes (Veletsos 1984). In this scenario, the pressure induced by

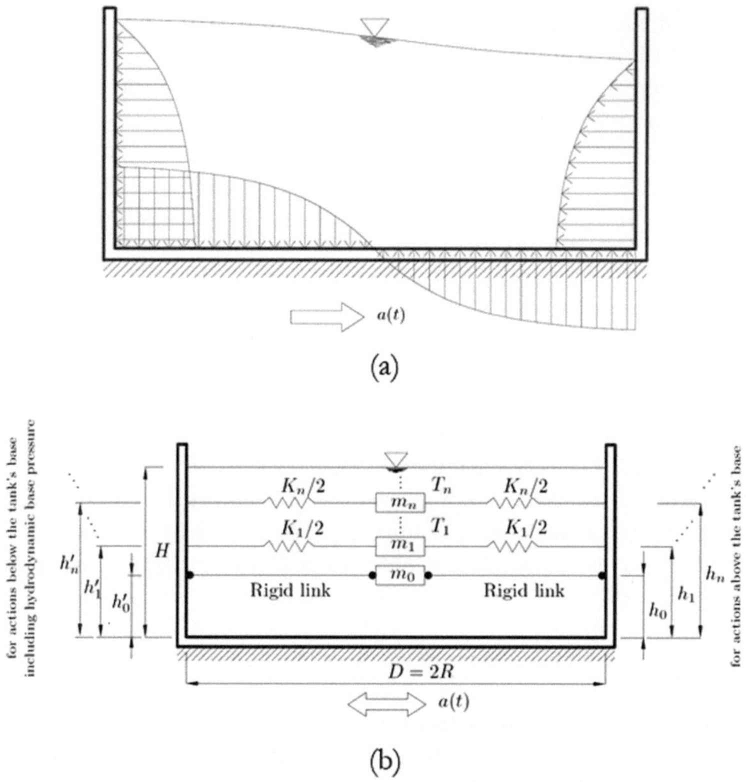

Fig. 2.2 **a** The hydrodynamic pressure distribution on the wall and base of a vertical tank caused by ground acceleration denoted as $a(t)$ is examined; **b** additionally, an equivalent mechanical model is introduced to describe the response of the contents in a rigid tank, distinguishing between impulsive and convective motions

the convective component is maximum at the liquid surface and is associated with low-frequency oscillations.

It is frequently beneficial to replace the liquid with an equivalent mechanical model conceptually. This model is based on a system of rigid bodies, spring masses, and stiffness (Ibrahim 2005). Pioneering work Dodge (1966) has led to the development of a straightforward mechanical analogy as depicted in Fig. 2.2b, specifically for the case of a rigid tank wall.

When the walls of the tank depicted in Fig. 2.2a undergo back-and-forth acceleration, a specific portion of the liquid, commonly referred to as impulsive mass (m_0 or m_i), is compelled to engage in this motion and imparts a reactive force onto the tank. The mass (m_0) is rigidly connected to the wall at a height (h_0 or h_i) ensuring that the horizontal force it generates aligns with the resultant force exerted by the equivalent impulsive fluid (Veletsos and Yang 1977). A comprehensive mechanical analogy for transverse sloshing necessitates incorporating an infinite number of oscillating masses ($m_1, m_2, \ldots, m_n, \ldots$), one for each of the infinite normal sloshing modes. These masses are interconnected to the tank walls through springs with stiffness (Chopra 1995):

$$K_i = \frac{4\pi^2 m_i}{T_i^2} \tag{2.1}$$

The flexibility of the attached masses corresponds to the different antisymmetric slosh frequencies of the fluid in the tank linked with vertical displacements. The combined mass, consisting of impulsive and convective masses ($m_0, m_1, m_2, \ldots, m_n, \ldots$), equals the total mass of the liquid (m).

The mass m_0 and lever arms h_0, h'_0 for impulsive rigid motion of a cylindrical vertical rigid tank may be calculated according to the following formulas:

For welded and bolted steel tanks as given by Formulas (2.2)–(2.7):

$$\frac{h'_0}{H} = \frac{3}{8}\left[\left(1 + \frac{4}{3}\left(\frac{\sqrt{3}\frac{R}{H}}{\tanh\left(\sqrt{3}\frac{R}{H}\right)} - 1\right)\right)\right] \quad \forall \frac{D}{H} \geq 1.333 \tag{2.2}$$

$$\frac{m_0}{m} = 1 - 0.218\frac{D}{H} \quad \forall \frac{D}{H} < 1.333 \tag{2.3}$$

$$\frac{h_0}{H} = \frac{3}{8} \quad \forall \frac{D}{H} \geq 1.333 \tag{2.4}$$

$$\frac{h_0}{H} = 0.5 - 0.094\frac{D}{H} \quad \forall \frac{D}{H} < 1.333 \tag{2.5}$$

$$\frac{h'_0}{H} = \frac{3}{8}\left[\left(1 + \frac{4}{3}\left(\frac{\sqrt{3}\frac{R}{H}}{\tanh\left(\sqrt{3}\frac{R}{H}\right)} - 1\right)\right)\right] \quad \forall \frac{D}{H} \geq 1.333 \tag{2.6}$$

$$\frac{h'_0}{H} = 0.5 + 0.06\frac{D}{H} \quad \forall \frac{D}{H} < 1.333 \tag{2.7}$$

For reinforced concrete or prestressed precast reinforced concrete tanks as given by Formulas (2.8)–(2.12):

$$\frac{m_0}{m} = \frac{\tanh\left(\sqrt{3}\frac{R}{H}\right)}{\sqrt{3}\frac{R}{H}} \quad \forall \frac{D}{H} \tag{2.8}$$

$$\frac{h_0}{H} = \frac{3}{8} \quad \forall \frac{D}{H} \geq 1.333 \tag{2.9}$$

$$\frac{h_0}{H} = 0.5 - 0.09375\frac{D}{H} \quad \forall\frac{D}{H} < 1.333 \tag{2.10}$$

$$\frac{h_0'}{H} = \frac{\sqrt{3}\frac{R}{H}}{2\tanh\left(\sqrt{3}\frac{R}{H}\right)} - \frac{1}{8} \quad \forall\frac{D}{H} \geq 0.75 \tag{2.11}$$

$$\frac{h_0'}{H} = 0.45 \quad \forall\frac{D}{H} < 0.75 \tag{2.12}$$

The mass m_1 and lever arms h_1, h_1' for the first-mode convective motion may be calculated according to the following formulas:

For welded and bolted steel tanks as given by Formulas (2.13)–(2.15):

$$\frac{m_1}{m} = \frac{1}{4}\sqrt{\frac{27}{8}\frac{R}{H}}\tanh\left(\sqrt{\frac{27}{8}\frac{H}{R}}\right) \tag{2.13}$$

$$\frac{h_1}{H} = 1 - \frac{\cosh\left(\sqrt{\frac{27}{8}\frac{H}{R}} - 1\right)}{\sqrt{\frac{27}{8}\frac{H}{R}}\sinh\left(\sqrt{\frac{27}{8}\frac{H}{R}}\right)} \tag{2.14}$$

$$\frac{h_1'}{H} = 1 - \frac{\cosh\left(\sqrt{\frac{27}{8}\frac{H}{R}} - \frac{31}{16}\right)}{\sqrt{\frac{27}{8}\frac{H}{R}}\sinh\left(\sqrt{\frac{27}{8}\frac{H}{R}}\right)} \tag{2.15}$$

For reinforced concrete or prestressed precast reinforced concrete tanks as given by Formulas (2.16)–(2.18):

$$\frac{m_1}{m} = \frac{1}{4}\sqrt{\frac{27}{8}\frac{R}{H}}\tanh\left(\sqrt{\frac{27}{8}\frac{H}{R}}\right) \tag{2.16}$$

$$\frac{h_1}{H} = 1 - \frac{\cosh\left(\sqrt{\frac{27}{8}\frac{H}{R}} - 1\right)}{\sqrt{\frac{27}{8}\frac{H}{R}}\sinh\left(\sqrt{\frac{27}{8}\frac{H}{R}}\right)} \tag{2.17}$$

$$\frac{h_1'}{H} = 1 - \frac{\cosh\left(\sqrt{\frac{27}{8}\frac{H}{R}} - 2.01\right)}{\sqrt{\frac{27}{8}\frac{H}{R}}\sinh\left(\sqrt{\frac{27}{8}\frac{H}{R}}\right)} \tag{2.18}$$

The first natural period T_c of the convective mode may be calculated using the following expression:

$$T_c = \frac{2\pi\sqrt{\frac{R}{g}}}{\sqrt{1.841\tanh\left(\frac{1.841H}{R}\right)}} \tag{2.19}$$

In case of rigid tanks without consideration of soil-structure interaction, the impulsive rigid period T_{ir} should be taken equal to zero. The impulsive rigid period T_{ir} of vibration of the tank-foundation system, including soil-structure interaction, may be calculated using Eq. (2.20):

$$T_{ir} = 2\pi\sqrt{\frac{m_0 + (m_b + m_w + m_r)}{K_x} + \frac{m_0 h_0^2}{K_\theta}} \tag{2.20}$$

where m_0 is the impulsive rigid mass; m_b, m_w, m_r are the base, wall, and roof masses; h_0 is the impulsive rigid lever arm without bottom pressure; K_x and K_θ are the horizontal and rocking stiffness of the foundation given in Sect. 2.5.

In case of rigid tank without consideration of soil-structure interaction, the vertical impulsive rigid period T_{irvert} should be taken equal to zero. The vertical period of vibration of the tank-foundation system, including soil-structure interaction, may be calculated using the following expression:

$$T_{irvert} = 2\pi\sqrt{\frac{m_V}{K_V}} \tag{2.21}$$

where m_V is the total mass of the tank and its liquid: $m_V = m + m_b + m_w + m_r$; m_b, m_w, m_r are the base, wall, and roof masses; K_V is the vertical stiffness of the foundation given in Sect. 2.5.

The Eurocode 8 Part 4 at section A.3.2.2 outlines a simplified procedure for fixed-base cylindrical tanks introduced in Malhotra (2000). The system comprising the tank and its liquid content is depicted using a two-degree-of-freedom model as illustrated in Fig. 2.3.

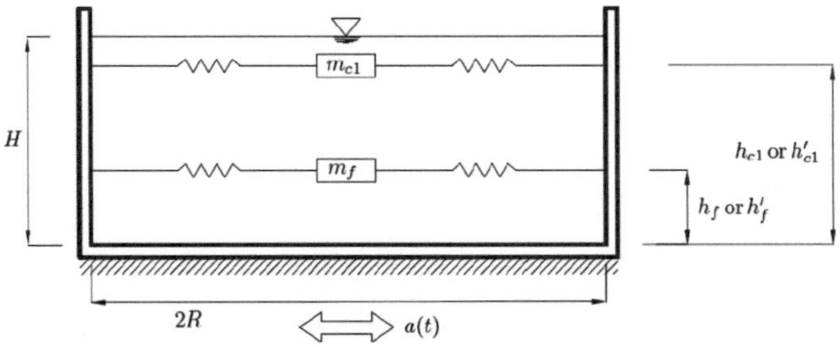

Fig. 2.3 Equivalent mechanical model representing the behavior of a deformable-wall tank filled with water and subjected to dynamic excitation: Malhotra's simplified procedure

The first degree of freedom corresponds to the impulsive component m_i which moves in solidarity with the tank wall, while the second is given by the convective component m_c. The natural periods of the impulsive and convective responses in seconds are:

$$T_{imp} = C_i \frac{H\sqrt{\rho}}{\sqrt{\frac{t}{R}}\sqrt{E}} \tag{2.22}$$

$$T_{con} = C_c\sqrt{R} \tag{2.23}$$

where H is the design height of the fluid in meters, R is the radius of the tank in meters, t is the thickness of the tank wall in meters (for tanks with non-uniform wall thickness t can be calculated by averaging the wetted height H of the tank wall), ρ is the mass density of the fluid in kg/m^3, and E is the modulus of elasticity of the tank material in N/m^2. The coefficients C_i and C_c are obtained from Fig. 2.4 (Section A.3.2.2.1 of Eurocode 8 Part 4): the C_i coefficient is dimensionless, while if R is in meters then C_c is expressed in s·m$^{-0.5}$. The impulsive and convective masses m_i and m_c are provided in Fig. 2.4 as functions of the total liquid mass m_i and the heights from the tank base of the points of application of the resultant impulsive and convective hydrodynamic pressures h_i and h_c are given as functions of the design height of the fluid.

The total shear at the base, the overturning moment immediately above and below the tank bottom (the overturning moment below the base plate that takes into account the hydrodynamic pressure both on the wall and the bottom of the tank) is calculated using the following expressions:

$$Q = (m_i + m_w + m_r)S_e(T_{imp}) + m_c S_e(T_{con}) \tag{2.24}$$

$$M = (m_i h_i + m_w h_w + m_r h_r)S_e(T_{imp}) + m_c h_c S_e(T_{con}) \tag{2.25}$$

$$M' = (m_i h_i' + m_w h_w + m_r h_r)S_e(T_{imp}) + m_c h_c' S_e(T_{con}) \tag{2.26}$$

where m_w is the mass of the tank wall and m_r is the mass of the tank roof; $S_e(T_{imp})$ is the impulsive spectral acceleration obtained from an elastic response spectrum

H/R	C_i	C_c	m_i/m_l	m_c/m_l	h_i/H	h_c/H	h'_i/H	h'_c/H
0.3	9.28	2.09	0.176	0.824	0.400	0.521	2.640	3.414
0.5	7.74	1.74	0.300	0.700	0.400	0.543	1.460	1.517
0.7	6.97	1.60	0.414	0.586	0.401	0.571	1.009	1.011
1	6.36	1.52	0.548	0.452	0.419	0.616	0.721	0.785
1.5	6.06	1.48	0.686	0.314	0.439	0.690	0.555	0.734
2	6.21	1.48	0.763	0.237	0.448	0.751	0.500	0.764
2.5	6.56	1.48	0.810	0.190	0.452	0.794	0.480	0.796
3	7.03	1.48	0.842	0.158	0.453	0.825	0.472	0.825

Fig. 2.4 Coefficients C_i and C_c, impulsive and convective masses, and impulsive and convective heights ($m_i, m_c, h_i, h_c, h_i', h_c'$) used in the Malhotra procedure

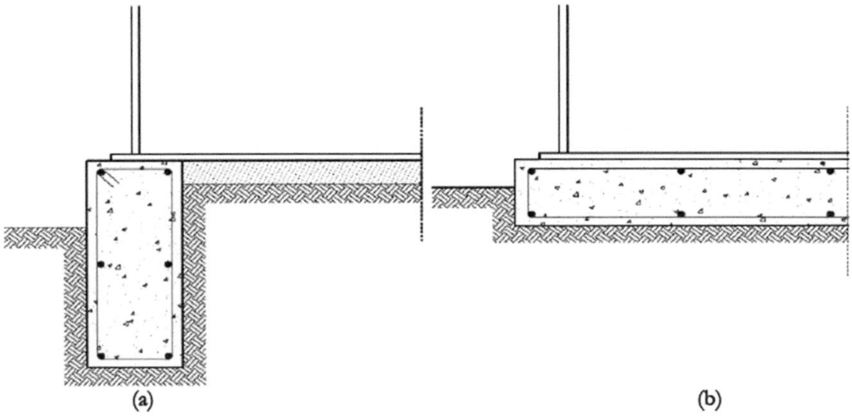

Fig. 2.5 Foundation for a circular steel tank: **a** ring foundation; **b** base plate foundation

with 2% damping for steel or pre-stressed concrete tanks, or from an elastic response spectrum with 5% damping for ordinary concrete tanks; $S_e(T_{con})$ is the convective spectral acceleration obtained from a damped elastic response spectrum with 0.5% damping; h_w and h_r are the heights of the centers of gravity of the tank wall and roof; h_i' and h_c' are the heights from the tank base of the points of application of the resultant impulsive and convective hydrodynamic pressures, taking into account the contribution of the pressure on both the wall and the bottom of the tank.

When the tank is equipped with a ring foundation (Fig. 2.5a), the moment M is used for the design of the wall base anchors and the foundation itself. If the tank is supported on a continuous base plate (Fig. 2.5b), M is employed for the design of the wall and anchors, and M' for the foundation.

2.5 Effects of Soil-Structure Interaction

The interplay between soil and structure significantly influences the response of tanks and silos to seismic events. Investigating the effects of soil-structure interaction involves assessing how the characteristics of the underlying soil impact the dynamic behavior of the equipment. This examination informs foundation design and retrofitting strategies, aiming to optimize the interaction between the structure and its supporting soil during seismic events. In case of a circular rigid foundation supported at the surface of a homogeneous soil deposit, the stiffnesses K_x and K_θ (Eq. 2.20) may be calculated as given in the following equations:

The horizontal stiffness K_x at the level of the tank's base may be calculated using Eq. (2.27):

$$K_x = \frac{8}{2 - \nu_s} G_s R_b \alpha_x \tag{2.27}$$

where: R_b is the radius of the foundation; G_s is the shear modulus of the soil; ν_s is the Poisson's ratio of the soil.

The rotational stiffness K_θ may be calculated by Eq. (2.28):

$$K_\theta = \frac{8}{3(1-\nu_s)} G_s R_b^3 \alpha_\theta \tag{2.28}$$

The dimensionless coefficients α_x and α_θ may be calculated as functions of the Poisson's ratio for the soil ν_s and a dimensionless frequency parameter as given by Formula (2.29):

$$\alpha = \frac{2\pi R_b}{T_{ir} \nu_s} \tag{2.29}$$

where: $\nu_s = \sqrt{\frac{G_s}{\rho_s}}$ is the shear wave velocity of the soil; ρ_s is the soil mass density.

The vertical stiffness K_V (Eq. 2.21) of the foundation should be calculated by Expression (2.30):

$$K_V = \frac{4}{1-\nu_s} G_s R_b \alpha_V \tag{2.30}$$

where: α_V is dimensionless coefficient obtained by Formula (2.29) with the period T_{ir} replaced by the vertical period $T_{ir,vert}$.

2.6 Base-Isolation Devices

Exploring the efficacy of base-isolation devices in mitigating seismic risks is integral to the overall seismic design strategy. These devices, designed to decouple the structure from ground motion, can be instrumental in reducing the transmission of seismic forces to the industrial equipment. An in-depth analysis of various base-isolation techniques and their applicability to tanks and silos will be conducted to determine their suitability for enhancing seismic resilience.

Using base isolation systems or damping systems in steel tanks can be instrumental in reducing vulnerability to seismic events. Here's an overview of their use, efficiency, advantages, and importance:

1. **Use**

   - Base isolation systems and damping systems are seismic retrofitting techniques designed to enhance the seismic performance of structures, including steel tanks.
   - These systems are employed to mitigate the impact of ground motion during an earthquake and reduce the forces transferred to the structure.

2. **Efficiency**:

   - Base isolation systems introduce flexibility at the base of the structure, allowing it to decouple from ground motion. This helps in isolating the tank from the full impact of seismic forces.
   - Damping systems, such as tuned mass dampers or viscous dampers, dissipate energy and reduce the amplitude of vibrations, preventing excessive swaying or rocking of the tank and also floating roof.

Table 2.1 provides a concise summary of the scientific literature pertaining to seismic isolated tanks reveals several notable isolation systems. These systems include Laminated Rubber Bearing (LRB), High Damping Rubber Bearing (HDRB), Lead-rubber bearing (NZ), Pure-Friction system (P-F), Friction Pendulum System (FPS), Variable Friction Pendulum System (VFPS), Electricite de France system (EDF), Vertical Isolation System (VIS), Sliding Concave Bearings (SCB), and Resilient-Friction Base Isolator (R-FBI). Each of these systems has been extensively studied and analyzed in the scientific literature to assess their efficacy in enhancing the seismic resilience of liquid storage tanks.

3. **Advantages**:

   - **Enhanced Seismic Performance**: Base isolation and damping systems improve the seismic resilience of steel tanks, minimizing damage and maintaining structural integrity during seismic events.
   - **Reduced Dynamic Forces**: These systems help in reducing the dynamic forces transmitted to the tank, preventing structural damage and ensuring the safety of the stored contents.
   - **Cost-Effective Retrofitting**: Retrofitting existing steel tanks with base isolation or damping systems can be a cost-effective way to enhance their seismic performance compared to rebuilding the entire structure.

4. **Importance**:

   - **Life Safety**: Implementing base isolation or damping systems is crucial for ensuring the safety of individuals working around or near the steel tank during an earthquake.
   - **Asset Protection**: These systems help protect the valuable contents of the tank, such as liquids or chemicals, by minimizing the risk of leakage or spillage during seismic events.
   - **Structural Resilience**: By reducing the vulnerability of steel tanks to seismic forces, these systems contribute to the overall resilience and longevity of the infrastructure.

Table 2.1 Summary of scientific literature on seismic isolation systems for tanks

Research	Type of tanks	Isolation systems	External action
Bo and Jia-xiang (1994)	Ground supported cylindrical steel	NZ	Unidirectional
Malhotra (1997a, b)	Ground supported cylindrical steel	Ring of vertically soft rubber bearing and ring of horizontally flexible bearing	Uni-bidirectional
Shenton III and Hampton (1999)	Elevated cylindrical steel	LRB	Unidirectional
Park et al. (2004a, b)	Pool-type rectangular concrete	HDRB	Unidirectional
Wang et al. (2001)	Ground supported cylindrical	FPS	Unidirectional
Kim et al. (2002)	Ground supported cylindrical	5 Horizontal $3.3 \cdot 10^5$ kg/m and 5 vertical $3.116 \cdot 10^5$ kg/m stiffness	Unidirectional
Shrimali and Jangid (2002a, b, c, 2003a, b, c, 2004)	Ground supported and elevated cylindrical steel	LRB, NZ, P-F, FPS, R-FBI, EDF	Uni-bidirectional
Cho et al. (2004)	Ground supported cylindrical steel	NZ	Unidirectional
Jadhav and Jangid (2004)	Ground supported cylindrical steel and concrete	HDRB and NZ	-
Gregoriou et al. (2005)	Ground supported cylindrical steel	FPS and VFPS	Unidirectional
Panchal and Jangid (2008)	Ground supported cylindrical steel	FPS and NZ	Unidirectional
Christovasilis and Whittaker (2008)	Ground supported rectangular concrete	FPS and NZ	Uni-bidirectional
Rawat et al. (2019)	Ground supported cylindrical steel	FPS and NZ	Bidirectional
Saha et al. (2013)	Ground supported cylindrical steel	FPS, NZ, VIS	Unidirectional and vertical
Kalantari et al. (2019)	Ground supported cylindrical steel	SCB	Unidirectional

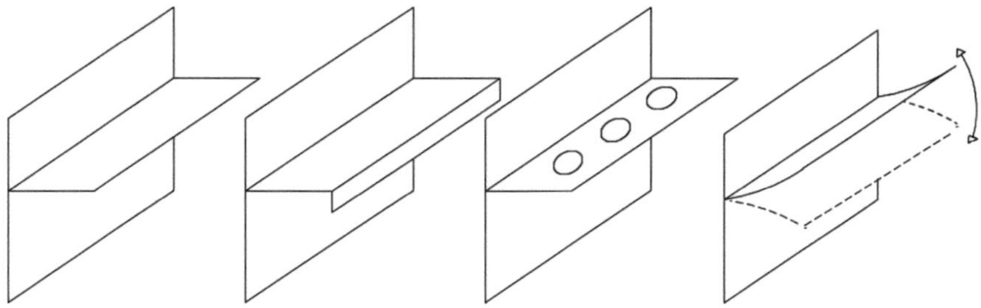

Fig. 2.6 Examples of rigid and flexible baffles

In summary, the use of base isolation and damping systems in steel tanks is a proactive approach to enhancing seismic resilience. It not only safeguards life and assets but also ensures the continued functionality of critical structures in seismic-prone areas.

Shrimali and Jangid conducted an insightful comparative study exploring the performance of various isolation systems for liquid storage tanks under real earthquake motions as detailed in their work (Shrimali and Jangid 2002a). Additionally, Table 2.1 provides a comprehensive overview of scientific research on the seismic isolation of fluid storage tanks as meticulously investigated by Panchal and Soni (2014). In the realm of sloshing behavior in tanks subjected to horizontal seismic excitation, Ma and Chang (1985), Maleki and Ziyaeifar (2007), along with Shekari et al. (2019) undertook a comprehensive study. Notably, these investigations considered the simultaneous effects of internal flow-dampening devices (Fig. 2.6) and external seismic isolations, contributing valuable insights to the understanding of tank dynamics under seismic conditions.

2.7 Some Final Considerations

In this chapter, a comprehensive and simplified exploration of the vulnerability of steel tanks in industrial plants has been undertaken, shedding light on the potential risks associated with seismic events. More insight can be found in Calvi and Nascimbene (2011). The critical role of steel tanks in storing hazardous materials and their susceptibility to structural damage during earthquakes underscore the paramount importance of understanding and mitigating this vulnerability.

Understanding Vulnerability: The analysis has revealed that the vulnerability of steel tanks is multifaceted, influenced by factors such as tank design, material properties, and seismic intensity. Recognizing the intricate interplay of these variables is pivotal in formulating effective protection measures.

Implications for Industrial Safety: The implications for industrial safety are profound. The structural integrity of steel tanks directly correlates with the prevention of catastrophic events, safeguarding both human lives and the environment. Addressing vulnerabilities becomes an ethical imperative in the context of industrial operations.

Protection Measures: Numerous protection measures have been explored in scientific literature, ranging from traditional seismic design principles to innovative technologies. Implementation of base isolation systems, damping devices, and structural retrofitting emerges as promising strategies. However, the selection of an appropriate protection measure necessitates a nuanced understanding of site-specific conditions and the criticality of the stored contents.

Recommendations for Future Research: While significant strides have been made in understanding vulnerability and devising protection measures, avenues for further research persist. Future investigations could delve into the dynamic behavior of steel tanks under different loading conditions, the cost-effectiveness of retrofitting strategies, and the integration of smart technologies for real-time monitoring.

Holistic Approach to Risk Mitigation: In conclusion, addressing the vulnerability of steel tanks demands a holistic approach. This involves integrating advanced engineering solutions with robust risk assessment methodologies. A proactive stance toward seismic resilience not only fortifies industrial infrastructure but also aligns with broader sustainability goals.

As we navigate the complexities of industrial safety, the insights gained from this chapter pave the way for informed decision-making and strategic interventions. By prioritizing the protection of steel tanks, we contribute to a resilient and secure industrial landscape that stands resilient in the face of seismic challenges.

References

Bo, L., & Jia-xiang, T. (1994). Vibration studies of base-isolated liquid storage tanks. *Computers & Structures, 52*(5), 1051–1059.

Calvi, G. M., & Nascimbene, R. (2023). *Seismic design and analysis of tanks*. Wiley. ISBN 978-1-119-84982-7.

Calvi, G. M., & Nascimbene., R. (2011) *Progettare i gusci*. IUSS Press.

Cho, K. H., Kim, M. K., Lim, Y. M., & Cho, S. Y. (2004). Seismic response of base isolated liquid storage tanks considering fluid structure soil interaction in time domain. *Soil Dynamics and Earthquake Engineering, 24*(11), 839–852.

Chopra, A. K. (1995). *Dynamics of structures: Theory and applications to earthquake engineering*. Prentice Hall.

Christovasilis, I. P., & Whittaker, A. S. (2008). Seismic analysis of conventional and isolated lng tanks using mechanical analogs. *Earthquake Spectra, 24*(3), 599–616.

Dodge, F. T. (1966). Analytical representation of lateral sloshing by equivalent mechanical models. In H. N. Abramson (Eds.), *The dynamic behaviour of liquids in moving containers* (pp. 198–223). NASA SP-106.

Gregoriou, V. P., Tsinopoulos, S. V., & Karabalis, D. L. (2005). Seismic analysis of base isolated liquefied natural gas tanks. In *Proceedings of the 5th GRACM International Congress on Computational Mechanics, Limassol, Cyprus* (pp. 305–312). GRACM.

Haroun, M. A. (1980). *Dynamic analyses of liquid storage tanks*. Technical report, California Institute of Technology.

Ibrahim, R. A. (2005). *Liquid sloshing dynamics: Theory and applications* Cambridge University Press.

Jadhav, M. B., & Jangid, R. S. (2004). Response of base-isolated liquid storage tanks. *Shock and Vibration, 11*, 276030.

Jaiswal, O. R., Rai, D. C., & Jain, S. K. (2007). Review of seismic codes on liquid containing tanks. *Earthquake Spectra, 23*(1), 239–260.

Kalantari, A., Nikoomanesh, M. R., & Goudarzi, M. A. (2019). Applicability of mass spring models for seismically isolated liquid storage tanks. *Journal of Earthquake and Tsunami, 13*(1), 1950002.

Kim, M. K., Lim, Y. M., Cho, S. Y., Cho, K. H., & Lee, K. W. (2002). Seismic analysis of base-isolated liquid storage tanks using the be-fe-be coupling technique. *Soil Dynamics and Earthquake Engineering, 22*(9–12), 1151–1158.

Ma, D. C., & Chang, Y. W. (1985). Analysis of seismic sloshing of reactor tanks considering submerged components and seismic isolation. In *ASME Pressure Vessels and Piping Conference* (pp. 98–7 139–147). American Society of Mechanical Engineers (ASME).

Maleki, A., & Ziyaeifar, M. (2007). Damping enhancement of seismic isolated cylindrical liquid storage tanks using baffles. *Engineering Structures, 29*(12), 3227–3240.

Malhotra, P. K. (1997). New method for seismic isolation of liquid-storage tanks. *Earthquake Engineering and Structural Dynamics, 26*(8), 839–847.

Malhotra, P. K. (1997). Method for seismic base isolation of liquid-storage tanks. *ASCE Journal of Structural Engineering, 123*(1), 113–116.

Malhotra, P. K. (2000). Practical nonlinear seismic analysis of tanks. *Earthquake Spectra, 16*(2), 473–492.

Panchal, V. R., & Jangid, R. S. (2008). Variable friction pendulum system for seismic isolation of liquid storage tanks. *Nuclear Engineering and Design, 238*(6), 1304–1315.

Panchal, V. R., & Soni, D. P. (2014). Seismic behaviour of isolated fluid storage tanks: A-state-of-the-art review. *KSCE Journal of Civil Engineering, 18*(4), 1097–1104.

Park, J.-H., Koh, H.-M., & Kim, J. K. (2004). Seismic isolation of pool-type tanks for the storage of nuclear spent fuel assemblies. *Nuclear Engineering and Design, 199*(1–2), 143–154.

Park, K. S., Koh, H. M., & Song, J. (2004). Cost-effectiveness analysis of seismically isolated pool structures for the storage of nuclear spent-fuel assemblies. *Nuclear Engineering and Design, 231*(3), 259–270.

Rawat, A., Matsagar, V. A., & Nagpal, A. K. (2019). Numerical study of base-isolated cylindrical liquid storage tanks using coupled acoustic-structural approach. *Soil Dynamics and Earthquake Engineering, 119*, 196–219.

Saha, S. K., Matsagar, V. A., & Jain, A. K. (2013). Comparison of base-isolated liquid storage tank models under bi-directional earthquakes. *Natural Science, 5*(8A1), 27–37.

Shekari, M. R., Hekmatzadeh, A. A., & Amiri, S. M. (2019). On the nonlinear dynamic analysis of base-isolated three-dimensional rectangular thin-walled steel tanks equipped with vertical baffle. *Thin-Walled Structures, 138*, 79–94.

Shenton, H. W., III., & Hampton, F. P. (1999). Seismic response of isolated elevated water tanks. *ASCE Journal of Structural Engineering, 125*(9), 965–976.

Shrimali, M. K., & Jangid, R. S. (2002). A comparative study of performance of various isolation systems for liquid storage tanks. *International Journal of Structural Stability and Dynamics, 2*(4), 573–591.

Shrimali, M. K., & Jangid, R. S. (2002). Earthquake response of liquid storage tanks with sliding systems. *Journal of Seismology and Earthquake Engineering, 4*(2 & 3), 51–61.

Shrimali, M. K., & Jangid, R. S. (2002). Non-linear seismic response of base-isolated liquid storage tanks to bi-directional excitation. *Nuclear Engineering and Design, 217*(1–2), 1–20.

Shrimali, M. K., & Jangid, R. S. (2003). Dynamic analysis of liquid storage tanks with sliding systems. *Advances in Structural Engineering, 6*(2), 145–158.

Shrimali, M. K., & Jangid, R. S. (2003). Earthquake response of isolated elevated liquid storage steel tanks. *Journal of Constructional Steel Research, 59*(10), 1267–1288.

Shrimali, M. K., & Jangid, R. S. (2003). Dynamic analysis of liquid storage tanks with sliding systems. *Journal of Seismology and Earthquake Engineering, 7*(3), 51–61.

Shrimali, M. K., & Jangid, R. S. (2004). Seismic response of base-isolated liquid storage tanks. *Journal of Sound and Vibration, 275*(1–2), 59–75.

Veletsos, A. S. (1984). Seismic response and design of liquid storage tanks. In *Guidelines for the seismic design of oil and gas pipeline systems* (pp. 255–370). ASCE Technical Council on Lifeline Earthquake Engineering.

Veletsos, A. S., & Yang, J. Y. (1977). Earthquake response of liquid-storage tanks. In *ASCE advanced in civil engineering through engineering mechanics* (pp. 1–24). North Carolina State University Raleigh.

Wang, Y.-P., Teng, M.-C., & Chung, K.-W. (2001). Seismic isolation of rigid cylindrical tanks using friction pendulum bearings. *Earthquake Engineering and Structural Dynamics, 30*(7), 1083–1099.

Chapter 3
Quantitative Risk Assessment of Natech Accidents

Federica Ricci and Valerio Cozzani

Abstract Natech (natural hazard-triggering technological) accidents are cascading events arising from the interaction between natural hazards and industrial installations, which pose unique challenges to risk management and emergency preparedness. These scenarios occur when natural events (including those of severe magnitude and low-intensity ones) impact industrial facilities, causing asset damage and triggering hazardous material releases. The occurrence of severe Natech accidents in the past has emphasized the need for comprehensive risk assessment methodologies and effective mitigation strategies. A critical component of Natech risk assessment is the evaluation of safety barriers, both technical (active and passive barriers) and procedural (e.g., emergency response). Indeed, the performance of safety barriers and safety systems may be compromised by the action of natural hazards, increasing the likelihood of technological scenarios. The degradation of safety barriers not only amplifies the risk but also increases the probability of domino effects, further amplifying the consequences of such accidents. This contribution presents methodologies available for Natech risk assessment, including quantitative approaches focusing on the quantitative assessment of safety barrier performance and domino effects in Natech scenarios, thus providing comprehensive and holistic risk figures for the management of Natech risk.

Keywords Natech · Chemical and process industry · Energy industry · Quantitative risk assessment · Safety barriers · Safety systems

3.1 Introduction

Natech accidents represent complex cascading events resulting from the interaction between natural and industrial hazards, presenting significant challenges for risk management and emergency preparedness (Salzano et al. 2013). These scenarios

F. Ricci · V. Cozzani (✉)
LISES—Laboratory of Industrial Safety and Environmental Sustainability, Department of Civil, Chemical, Environmental and Material Engineering, Alma Mater Studiorum—University of Bologna, Bologna, Italy
e-mail: valerio.cozzani@unibo.it

V. Cozzani and F. Paolacci (eds.), *Management Strategies and Resilience Against NaTech Events*, CISM International Centre for Mechanical Sciences 616,
https://doi.org/10.1007/978-3-031-84485-0_3

occur when natural disasters, such as earthquakes, floods, or storms, impact industrial sites, leading to the release of hazardous materials (Showalter and Myers 1994). The occurrence of severe Natech accidents, as those triggered by the Kocaeli earthquake (Girgin 2011a; Cruz and Steinberg 2005a) and the Great East Japan earthquake and tsunami (Krausmann and Cruz 2013a), has drawn the attention of regulatory authorities and plant operators to this category of accident scenarios. Moreover, the consequences of Natech accidents can be severe, posing threats to human life, the environment (Ricci et al. 2022b), and asset integrity (Ricci et al. 2021; Caratozzolo et al. 2022).

The development of robust methods for Natech risk assessment is paramount for gaining a holistic understanding of potential hazards and vulnerabilities, thereby supporting effective risk management strategies (Mesa-Gómez et al. 2021). Actually, a well-designed risk assessment methodology offers a structured framework for evaluating the likelihood and potential consequences of adverse events, facilitating informed decision-making and resource allocation (Center for Chemical Process Safety 2000; Uijt de Haag and Ale 2005; Mannan 2005).

The assessment and mitigation of the risks associated with Natech accidents demand a comprehensive assessment of various multidisciplinary factors, including the characterization of natural hazards (Selva et al. 2021), the quantification of possible damages to equipment items (Salzano et al. 2009; Landucci et al. 2012), and the assessment of potential cascading effects (Krausmann et al. 2010a). Moreover, the increasing frequencies and intensities of extreme natural events, at least in part attributable to climate change (Centre for Research on the Epidemiology of Disasters 2020), underscore the urgency of proactive risk management strategies for Natech accidents.

The implementation of safety barriers has always been one of the most widely adopted strategies to mitigate the frequency and reduce the consequences of technological accidents (Center for Chemical Process Safety 2001; Andersen et al. 2004). However, several case histories have highlighted the potential unavailability of conventional safety barriers and safety systems during Natech events (Girgin 2011a; Organisation for Economic Cooperation and Development 2015; U.S. Chemical Safety and Hazard Investigation Board 2018; Steinberg and Cruz 2004a; Misuri et al. 2019). This stems from the unique conditions caused by the impact of natural events. Actually, besides technical barriers, emergency response can be also affected by natural disasters. In fact, natural events may hinder and/or delay emergency response efforts, placing additional strain on emergency systems that must simultaneously address the consequences of both natural and technological disasters (Salzano et al. 2013; Krausmann and Salzano 2017; Lindell and Perry 1996; Ricci et al. 2022a). Criticalities related to emergency response become even more evident due to the increased probability of cascading events in Natech accidents compared to conventional scenarios. The inclusion of the possible degradation of safety systems in Natech quantitative risk assessment (QRA) is therefore paramount to provide correct risk figures.

In this chapter, an approach to the QRA of Natech accident scenarios is presented. The method also addresses the impact of safety barrier degradation on the frequency and severity of primary technological scenarios in Natech events. The approach incorporates the effects of degraded safety barriers on the overall risk metrics, considering the potential for unmitigated technological scenarios resulting from safety barrier failure, unavailability, or reduced performance during Natech events. In addition, methods available for the quantitative risk assessment of domino effects triggered by Natech accidents are illustrated, together with approaches for emergency response performance quantification.

3.2 Quantitative Risk Assessment of Primary Natech Scenarios

Figure 3.1 presents a general framework for Natech QRA that includes the assessment of the possible degradation of safety systems due to the effect of natural hazards. The methodology derives from the approach suggested by Cozzani et al. (2014) and Misuri et al. (2023). This method builds on specialized procedures designed for managing accidents triggered by earthquakes, floods, and lightning strikes (Cozzani et al. 2014; Antonioni et al. 2007, 2009a, 2015; Necci et al. 2016), which are based on the extension and modification of traditional risk assessment frameworks (Center for Chemical Process Safety 2000; Mannan 2005). The approach integrates the necessary steps for Natech risk assessment with those used in conventional accident scenarios, thereby streamlining the process and facilitating the comparison of Natech risk with that deriving from conventional accidents.

The following sections provide a detailed description of each step of the methodology, along with the required information to support risk quantification.

3.2.1 Definition and Characterization of Reference Natural Hazards

The initial phase of Natech Risk Assessment involves a quantitative analysis of the natural hazards that could affect a specific site (as outlined in Step 1 of Fig. 3.1). This step aims to identify and characterize the potential natural hazards at the site by determining their expected frequency (f_{nh}) or return period and magnitude, providing the input data needed for further analysis.

In the specific framework of Natech risk assessment, the magnitude or intensity of natural occurrences can be represented by an impact vector. This vector encompasses a spectrum of one or more intensity parameters, providing a simplified assessment of the severity of the natural hazard considered in relation to its return period. Table 3.1 provides a selective overview of the parameters used to define the intensity of various natural hazard categories.

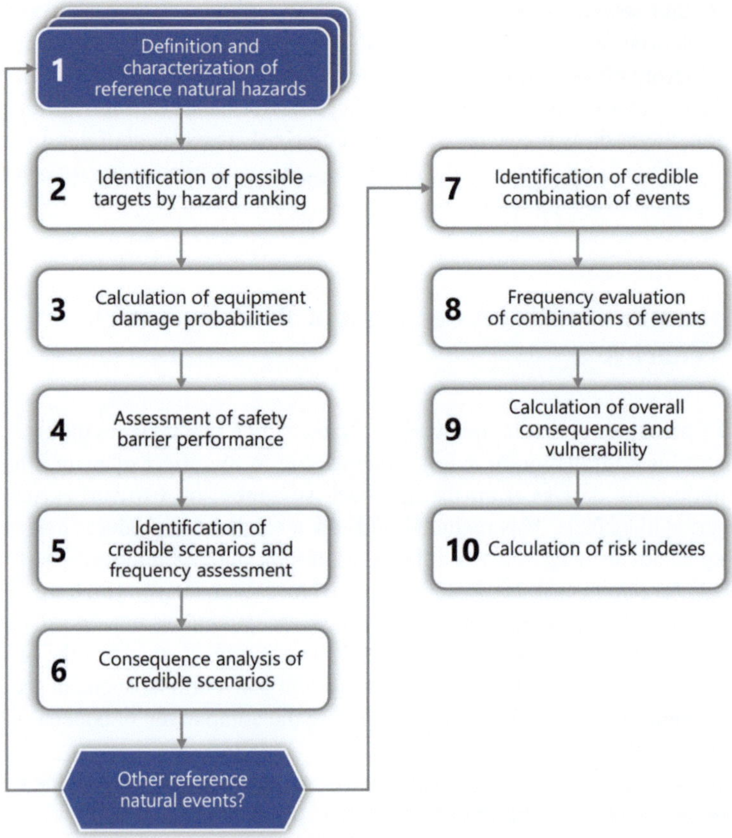

Fig. 3.1 General framework for Natech Quantitative Risk Assessment. Adapted from Misuri et al. (2023)

The characterization of natural hazards is a complex task, and a tailored approach to its quantification should be selected, based on the natural hazard considered. For instance, the intensity of earthquakes can be performed using a Probabilistic Seismic Hazard Analysis (PSHA), which allows the estimation of specific ground motion intensity measures for each component (Iervolino et al. 2014, 2016, 2015a). As in the case of tsunami hazards, the intensity evaluation can be performed through detailed simulations, which require extensive computational resources (Selva 2013; Selva et al. 2016; Mignan et al. 2014). Specialized filtering methods, as cluster analysis, can be used, building upon the outcomes of regional hazard assessments (Lorito et al. 2014). When dealing with floods, the flow field is typically simulated using depth-averaged Shallow Water Equations (SWE) (Aureli et al. 2015; Pilotti et al. 2011, 2010). Nonetheless, significant uncertainties persist, impacting both the river flood return period and intensity, related to the possible breaches or failure of

Table 3.1 Overview of the parameters used to characterize the magnitude of natural hazards

Natural hazard	Intensity parameters
Earthquake	Peak ground acceleration (g)
	Peak ground velocity ($m \cdot s^{-1}$)
Flood	Water depth (m)
	Water velocity ($m \cdot s^{-1}$)
Hurricane	Wind speed ($m \cdot s^{-1}$)
	Water velocity ($m \cdot s^{-1}$)
	Inundation depth (m)
	Accumulated rainwater (m)
Lightning	Flash density at ground level, n_g ($km^{-2} \cdot y^{-1}$)
Tsunami	Wave speed ($m \cdot s^{-1}$)
	Wave height [m]
Wildfire	Flame emissive power, E_{wf} ($W \cdot m^{-2}$)
	Flame length, L_f (m)
	Flame duration (s)
Wind	Wind speed ($m \cdot s^{-1}$)

levees. In the case of wildfires, existing models primarily focus on assessing affected areas and wildfire progression rather than predicting wildfire frequency, intensity, and duration at specific sites (Naderpour et al. 2019; Finney 1998; Tymstra et al. 2010; Hilton et al. 2014), thus not being directly applicable in the context of Natech QRA. These sophisticated approaches require detailed modeling, specific inputs, and high computational costs, which might not be applicable in the context of Natech QRA. As an alternative, simplified approaches based on hazard maps are also available, which provide a simplified characterization of the natural hazards suitable for application in the context of Natech assessment. In particular, hazard maps exist for specific natural events as earthquakes (Musson 2000; Woessner et al. 2015), floods (de Moel et al. 2009) and lightning (Cecil et al. 2014; International Electrotechnical Commission 2019, 2006). Moreover, generic hazard maps providing information on multiple hazards are available in the literature (Centro Internacional de Métodos Numéricos en Ingeniería et al. 2013; MunichRE 2011; Global Facility for Disaster Reduction and Recovery 2024; National Aeronautics and Space Administration 2024).

3.2.2 Identification of Possible Targets by Hazard Ranking

The identification of the more critical equipment items to be included in a Natech QRA is of paramount importance. In fact, including all equipment across an extensive industrial site in the QRA process could potentially burden the calculation procedure without enhancing the accuracy of the results. Therefore, it is important to implement

	Liquefied gas	Overheated liquid	Gas	Cryogenic liquid	Liquid
Storage vessels	4	3	3	2	1
Large diameter pipes	4	3	2	2	1
Columns	3	2	2	2	1
Process equipment	3	2	1	1	1

Fig. 3.2 Matrix for the identification of critical equipment items storing or handling flammable substances in quantitative risk assessment of Natech scenarios. (1: low criticality–4: high criticality). Adapted from Antonioni et al. (2009a) and Cozzani et al. (2014)

a preliminary screening procedure, akin to conventional risk assessment practices, to narrow down the analysis to a selected number of items recognized as the primary contributors to risk figures (Center for Chemical Process Safety 2000; Uijt de Haag and Ale 2005). When identifying critical equipment, several factors come into play, including the equipment type, hazardous substance properties, inventory, and operating conditions such as pressure and temperature. Various methods available in the technical literature can be used to carry out the screening of equipment items for inclusion in QRA studies (Center for Chemical Process Safety 2000). Among others, the screening matrixes proposed by Antonioni et al. (2009a) and Cozzani et al. (2014) are a valuable tool, since they allow considering three main criteria: the hazardous properties of the substance, the type of equipment items, and the physical state of the substance. An example of a matrix is reported in Fig. 3.2, specifically developed for the case of flammable substances.

Natech-specific factors should also be taken into account. In particular, it is important to analyze past accidents to identify the items that have been more frequently involved in past Natech events. Numerous studies in the literature have delved into past accidents, facilitating the identification of equipment most susceptible to damage (Cozzani et al. 2010; Krausmann et al. 2011b; Renni et al. 2010). For instance, Antonioni et al. (2009a) analyzed past accidents triggered by earthquakes and floods, determining that atmospheric tanks are frequently affected. Similarly, a study by Renni et al. (2010) shows that atmospheric tanks are among the more vulnerable items also in the case of lightning.

3.2.3 Calculation of Equipment Damage Probabilities

The calculation of the expected probability of damage resulting from the impact of a natural hazard (P_{nhd}) is typically based on simplified methods as observational fragility models or probabilistic equipment vulnerability models. These models are specifically developed for application within a Quantitative Risk Assessment (QRA) framework, necessitating minimal input data and limited computational resources.

Equipment damage models enable the estimation of an equipment damage probability based on one or more intensity parameters of the reference natural hazard, along with a few geometrical and operational parameters (e.g., diameter, height, filling level, etc.). The intensity parameters of the natural hazard are typically derived from the analysis carried out in the initial step of the procedure (Step 1 in Fig. 3.1). Table 3.2 provides a list of damage models available in the literature, typically used in the Natech QRA frameworks. Further details on the origins and application of these models can be found in specific publications (Cozzani and Salzano 2017).

The equipment damage models listed in Table 3.2 are specific for different categories of natural hazards and equipment types. The input parameters, assumptions and simplifications incorporated in the model, and the failure mode considered are also specific. Notably, two primary approaches were used for the development of fragility and damage models: (i) analysis of observational data; and (ii) the development of simplified mechanical models addressing specific equipment failure modes.

In addition to assessing the failure probability of equipment items, evaluating the intensity of Loss Of Containment (LOC) is crucial. LOC intensity is primarily influenced by the severity of structural damage, the physical state of the released substance, and operational parameters of the damaged vessel, such as operating pressure and temperature. Risk assessment frameworks recommend a simplified approach based on discrete damage states (DS), which stem from the failure mode considered by the equipment damage model. Widely applied definitions of damage states are provided by O'Rourke and So (2000) and HAZUS (1997) in the case of seismic damage. Additionally, discrete release states (RS) may be used to define the quantity of hazardous substances released, often based on the extent of equipment damage. For instance, in the case of seismic events, three levels of LOC intensity are typically considered (Salzano et al. 2009, 2003; Fabbrocino et al. 2005): no loss (RS1), moderate loss (RS2), and extensive loss of containment (RS3). Based on these release states, various release modes and their associated diameters can be defined (e.g., 10 mm release, complete inventory release over 10 min, instantaneous release of the complete inventory), and conventional source term models can be applied (Center for Chemical Process Safety 2000; Uijt de Haag and Ale 2005; Mannan 2005). Table 3.3 provides examples of damage states and release states adopted in literature studies conducting quantitative risk assessments of accidents triggered by earthquakes and floods.

Table 3.2 List of equipment damage models available in the literature for the calculation of the failure probability of equipment items due to natural hazards

Natural hazard	Equipment type	References
Earthquake	Vertical atmospheric storage tanks	Salzano et al. (2003, 2009)
	Horizontal storage vessels	Salzano et al. (2009)
	Pressurized reactors	Salzano et al. (2009)
	Pumps	Salzano et al. (2009)
	Pipelines	Lanzano et al. (2013)
	Underground pipelines	Lanzano et al. (2014)
	Spherical pressurized vessel	Moschonas et al. (2014)
Flood	Vertical atmospheric storage tanks	Landucci et al. (2012), Khakzad and Van Gelder (2017; 2018), Zuluaga Mayorga et al. (2019), Kameshwar and Padgett (2018; 2015)
	Horizontal storage vessels	Caratozzolo et al. (2022), Landucci et al. (2014)
	Horizontal process equipment	Caratozzolo et al. (2022)
	Pipeline	Rossi et al. (2022)
Hurricane	Vertical atmospheric storage tanks	Qin et al. (2020)
Lightning	All	Necci et al. (2013; 2014)
Tsunami	Vertical atmospheric storage tanks	Basco and Salzano (2017)
	Horizontal storage vessels	Basco and Salzano (2017)
Wind	Vertical atmospheric storage tanks	Ramírez Olivar et al. (2020)

3.2.4 Assessment of Safety Barrier Performance

In order to prevent or mitigate technological accidents, safety protocols and specialized technical measures, known as safety barriers, can be implemented. These barriers encompass various solutions such as water deluge systems to protect critical equipment from fire and catch basins to contain liquid spills. Past Natech events have demonstrated that conventional safety systems may prove inadequate during Natech scenarios due to the specific conditions resulting from the interaction with natural phenomena. Moreover, in several Natech events safety barriers and safety systems were affected as well by the natural hazard (Girgin 2011a; Organisation for Economic Cooperation and Development 2015; U.S. Chemical Safety and Hazard Investigation Board 2018; Steinberg and Cruz 2004a; Misuri et al. 2019). The degradation of safety barrier performance can lead to an increase in the frequency and severity of techno-

Table 3.3 Examples of damage states and related release states used in literature studies to carry out a Natech quantitative risk assessment

Equipment type	Natural hazard	Damage state	Release state	References
Atmospheric tanks	Earthquake	Minor structural damage	Not considered	Campedel et al. (2008a)
		Relevant structural damage	Leak with an equivalent diameter of 50 mm	
		Total collapse of the structure	Instantaneous loss of containment of the entire vessel inventory	
	Earthquake	Limited structural damage, as the rupture of connections or the buckling of equipment	Partial loss of vessel inventory or the entire loss in a time interval higher than 10 min	Antonioni et al. (2009a)
		Extended structural damage	Complete loss of containment of vessel inventory in a time interval lower than 10 min	
	Flood	Catastrophic failure and impact of/with adjacent vessels or with trailed objects	Instantaneous release of the complete inventory (in <2 min)	Antonioni et al. (2009a)
		Failure of roof or shell fracture	Continuous release of the complete inventory (in more than 10 min)	
		Failure of flanges and connections	Continuous release from a hole having an equivalent diameter of 10 mm	

(*continued*)

Table 3.3 (*continued*)

Equipment type	Natural hazard	Damage state	Release state	References
Pressurized tanks	Earthquake	Minor structural damage	Not considered	Antonioni et al. (2009a)
		Relevant structural damage	Release of the entire inventory in 10 min	
		Total collapse of the structure	Quasi-instantaneous release of the entire vessel inventory	
	Flood	Impact with/of adjacent vessels	Instantaneous release of the complete inventory (in <2 min)	(Antonioni et al. 2009a)
		Shell fracture	Continuous release of the complete inventory (in more than 10 min)	
		Failure of flanges and connections	Continuous release from a hole having an equivalent diameter of 10 mm	

logical scenarios, exacerbating both primary incidents and their escalation (Misuri et al. 2021b, c, 2023). Therefore, the inclusion of an ad-hoc assessment of safety barrier performance in Natech QRA is essential to provide accurate risk figures.

The assessment of technical safety barrier performance in Natech accidents may follow a two-step methodology, which is reported in Fig. 3.3. The approach applies to active and passive barriers. The definitions of these categories of safety barriers and safety systems are reported in Table 3.4.

In the first step, safety barriers need to be classified according to the definitions reported in Table 3.4. Then, the baseline performance is evaluated using two parameters: the Probability of Failure on Demand (PFD_0) and the effectiveness value (η_0). The PFD refers to the likelihood of the system being unavailable when needed, while η accounts for the probability of the barrier successfully executing its function once activated. For basic components, PFD values can typically be found in the technical literature (Mannan 2005; Veritas 2002; Cadwallader 1996; Institute 2016a, b; Commission 2008; Madonna et al. 2009). However, for complex systems as active safety barriers, determining PFD values often involves conducting a fault tree analysis

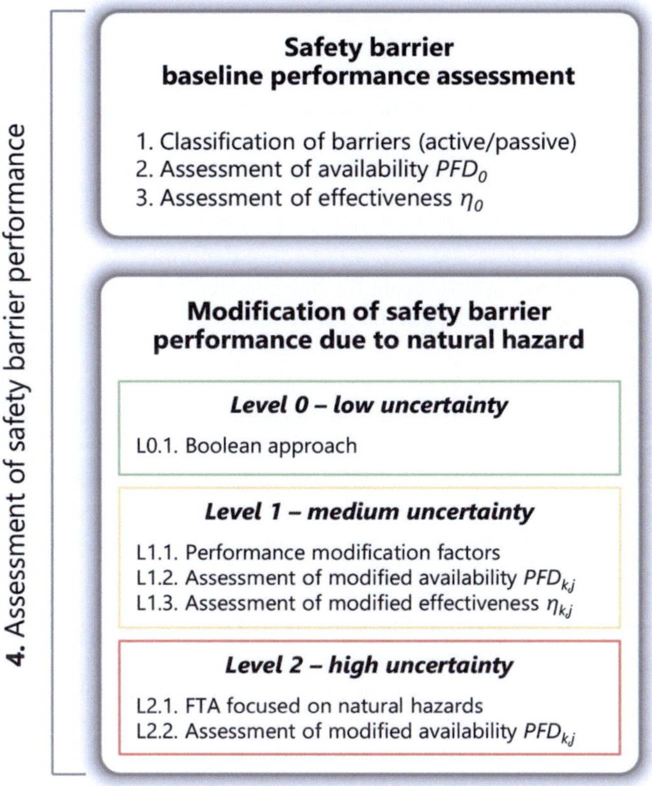

Fig. 3.3 Methodology for the assessment of safety barrier performance in Natech scenarios

Table 3.4 Definition of safety barriers category and examples

Barrier category	Definition	Examples
Passive	Physical technical systems that are permanently available and not require external activation to perform their function	Pressure safety valves, containment dikes, bunds, catch basins, fireproofing
Active	Technical barriers that require activation (e.g., automatic through a detection-processing-actuation loop, by operators) to perform their function	Sprinklers, water deluge systems, water curtains

Adapted from Misuri et al. (2023, 2020b)

(FTA) considering the system architecture. Alternatively, if data is scarce, simplified risk-based methods can be utilized to estimate *PFD* (Commission 2010, 2016). Determining the effectiveness parameter necessitates considering performance data specific to the type of system, as well as factors such as operational management, system installation, and maintenance (Landucci et al. 2015, 2017, 2016; Frank et al. 2013; Roberts 2004; Landucci et al. 2009).

The performance modification of safety systems due to the effects of natural hazards needs then to be assessed. Three methods are proposed for this aim, in order to appropriately consider barriers with increasing complexity:

- Level 0 (L0, low uncertainty) involves a simplified evaluation suitable for simpler barrier systems with low uncertainty regarding the impact of natural hazards. A Boolean approach is applied, using the baseline performance value if the barrier remains unaffected while considering the system unavailable if certainly impacted by the natural event.
- Level 1 (L1, intermediate uncertainty) applies performance modification factors for a broad set of safety barriers considering a reference architecture. It is suitable when some uncertainty regarding barrier performance is present.
- Level 2 (L2, high uncertainty) entails a detailed analysis of barrier architecture and subsystems, requiring the consideration of site-specific scenarios and special design provisions. It relies on an ad-hoc approach based on fault tree analysis (FTA) for complex active barrier systems.

The selection of the level of the analysis to be implemented depends on the uncertainty regarding the potential interaction between the reference natural hazard and the specific features of the barrier under consideration. Table 3.5 outlines the equation for evaluating modified safety barrier performance in the case of natural events according to the level of analysis considered.

The performance modification factors (φ) to be applied in the L1 approach were derived from expert judgments using a covariate method (Misuri et al. 2020b). In assessing the safety barriers, the second quartile of the failure probability distributions is employed to minimize outlier impacts. The performance modification factor can be interpreted as the probability that barriers are negatively affected or damaged by natural hazards, with higher values (approaching 1) suggesting a greater likelihood of barrier failure. Values of φ for a set of safety barriers exposed to floods and earthquakes are provided in Table 3.6, derived from the study by Misuri et al. (2020b). Additional safety systems can be found in the original study, together with performance distributions.

3.2.5 Identification of Credible Scenarios and Frequency Assessment

The integration of safety barriers into Natech quantitative risk assessment necessitates a shift from the binary perspective, where an item is either completely safe or involved in an accident. Indeed, it is crucial to consider scenarios where outcomes are mitigated due to the influence of safety barriers and safety systems. This requires

Table 3.5 Approach to assess safety barrier performance in the case of natural events according to the level of analysis considered

Level	Performance modification	
L0	Active barriers	$PFD_{j,k} = 1 + \left(\gamma_{j,k}) - 1\right) \cdot \left(1 - PFD_{j,k}\right)$
		$\eta_{j,k} = \eta_{0,k}$
	Passive barriers	$\eta_{j,k} = \left(1 - \gamma_{j,k}\right) \cdot \eta_{0,k}$
	Where $\gamma_{j,k}$ is equal to 1 if the *k*-th barrier is deemed affected by the *j*-th natural event, 0 otherwise	
L1	Active barriers	$PFD_{j,k} = 1 + \left(\varphi_{j,k} - 1\right) \cdot \left(1 - PFD_{j,k}\right)$
		$\eta_{j,k} = \eta_{0,k}$
	Passive barriers	$\eta_{j,k} = \left(1 - \varphi_{j,k}\right) \cdot \eta_{0,k}$
	Where $\varphi_{j,k}$ is the performance modification factor of the *k-th* barrier considering the *j-th* natural event. Examples of factors are reported in Table 3.6	
L2	Active barriers	$PFD_{j,k} = 1 - \prod_m \left(1 - Q_j\left(MCS_{m,k}\right)\right)$
		$Q_j\left(MCS_{m,k}\right) = \prod_p \left(q_{p,0} + \delta_{p,j}\left(1 - q_{p,0}\right)\right)$
		$\eta_{j,k} = \eta_{0,k}$
	Where $Q_j\left(MCS_{m,k}\right)$ represent the unavailability of the *m*-th minimal cut set (*MCS*) identified from the FTA, $q_{p,0}$ is the probability of the *p*-th basic event in the *m*-th *MCS*, and $\delta_{p,j}$ takes into account the vulnerability of the *p*-th basic event to the *j*-th natural event ($\delta_{p,j} = 1$ if the *p*-th event is deemed affected, 0 otherwise)	

Adapted from Misuri et al. (2023, 2021c)

Table 3.6 Performance modification factors for some safety barriers in case of earthquakes φ_e and floods φ_f

Safety barrier	φ_e	φ_f
Automatic rim-seal fire extinguishers	0.5	0.15
Fixed/semi-fixed foam systems	0.5	0.375
WDS/water curtains/sprinklers	0.75	0.375
Fire and gas detectors	0.5	0.5
Shutdown valves	0.5	0.25
Fire walls	0.5	0.2
Fireproofing	0.25	0.15
Bunds/catch basins	0.5	0.75
Burying tanks	0.25	0.15

Data from Misuri et al. (2020b). Additional safety systems can be found in the original study, together with performance distributions

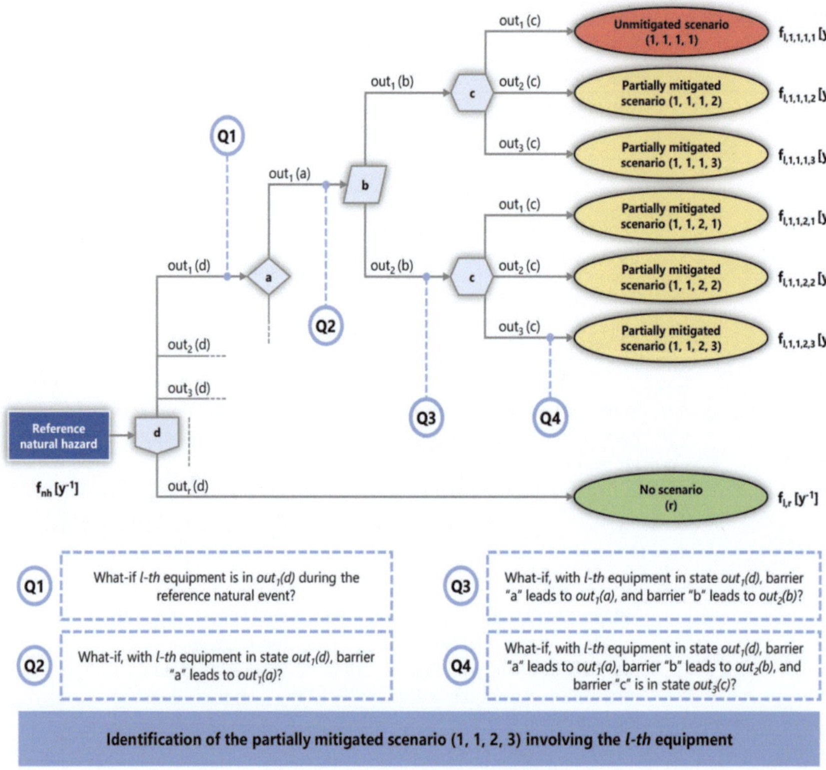

Fig. 3.4 Generic event tree developed for the identification and assessment of unmitigated and partially mitigated scenarios considering the impact of a reference natural hazard on the generic *l*-th equipment item including the results of what-if analysis (blue-dashed boxes). Adapted from Misuri et al. (2023)

a specific methodology to identify and assess the frequency of the expected scenarios (as outlined in Step 5 of Fig. 3.1) and to evaluate their consequences (Step 6 of Fig. 3.1) considering the potential mitigation due to safety barriers.

The inclusion of safety barriers in the QRA thus requires considering their effects by defining a range of potential final outcomes. These may vary on the basis of the barriers implemented, their safety functions, and their performance. When multiple safety barriers are linked to a single piece of equipment, a spectrum of unmitigated and partially mitigated outcomes needs to be taken into account, depending on the activation and/or effectiveness of each mitigative safety barrier. The identification and frequency assessment of these outcomes are obtained by the development of a specific event tree using "What-If" analysis (Mannan 2005). Event tree analysis (ETA) can be used to determine the conditional probabilities of these outcomes. An example of an event tree is illustrated in Fig. 3.4, and the logical operators involved are detailed in Table 3.7.

The process begins with characterizing the possible Loss of Containment (LOC) scenarios for a generic lth piece of equipment during an extreme natural event, as depicted in Fig. 3.4. Gate "d" incorporates all potential states of the l-th equipment during the event (see Table 3.7), representing all failure modes leading to LOCs and the possibility of no failure. While some vulnerability models only account for two states, either failure with LOC or complete safety, others include multiple risk states. For instance, the vulnerability model for vertical atmospheric tanks facing flood conditions developed by Landucci et al. (2012) addresses only buckling failure, providing a failure probability (P_{nhd}) for this mode. In contrast, several fragility models for equipment items subjected to seismic loads consider various damage and release states, each with a distinct P_{nhd}, as those provided by Salzano et al. (2009). The possible $(r - 1)$ values of P_{nhd} have been estimated in Step 3 (see Sect. 3.2.3), and are subsequently utilized to quantify the "d" gates in the ETA.

The characterization of outcomes, both unmitigated and partially mitigated, following intermediate events using the"What if?" approach (Mannan 2005) is demonstrated through an example in Fig. 3.4. Specifically, the figure illustrates the identification of a partially mitigated scenario resulting from the l-th target in state '1', and involves three safety barriers in states '1', '2', and '3'. This method is replicated for each expected state of the l-th equipment and of safety barriers, identifying numerous potential primary scenarios as indicated on the right side of Fig. 3.4.

A top-down approach is recommended, beginning with expected Loss of Containment (LOC) events to outline intermediate events in the Event Tree Analysis (ETA). These events hinge on the distinct safety functions and performance of each barrier. Integration of barriers into the ETA employs logical gates as shown in Table 3.7, adhering to logical rules consistent with those in previous studies (Misuri et al. 2021c; Landucci et al. 2015, 2016). The ETA supports the evaluation of the frequency of each of the total m_l primary scenarios for the l-th equipment item, starting from f_{nh} values assessed in Step 1 of Fig. 3.1 (see Sect. 3.2.1), incorporating the damage probability P_{nhd} calculated in Step 3 (see Sect. 3.2.3) and performance updates for each barrier expressed as $PFD_{j,k}$ and $\eta_{j,k}$ values, determined in Step 4 (see Sect. 3.2.4).

In addition to safety barriers and safety systems, the construction of the event tree should encompass standard gates commonly incorporated in traditional Event Tree Analysis. For instance, when evaluating flammable substances, it is imperative to contemplate both immediate ignition, delayed ignition, or the possibility of flash fire and vapor cloud explosion within the analysis. It is important to consider that the probabilities of ignition of substances can also be influenced by the effects of natural hazards. Indeed, specific conditions may occur that reduce the likelihood of ignition, as in the case of floods, or increase its probability, as in the case of lightning. Ricci et al. (2021) devised quantitative post-release event trees tailored to distinct substance categories and classes of natural hazards (geophysical, hydrogeological, and meteorological). The quantification process drew from an examination of over 6400 historical accident records. Reference ignition probabilities associated with specific natural hazards were obtained (see Table 3.8).

Figure 3.5 illustrates a qualitative example of event trees depicting primary Natech scenarios involving a pressurized vessel exposed to an earthquake event, with a

Table 3.7 Operators defined for the ETA of primary Natech scenarios considering the presence of safety barriers. fin: gate input frequency. f_{nh}: reference natural hazard frequency. *PFD*: probability of failure on demand. η: effectiveness parameter. P_{nhd}: equipment damage probability of one of the $(r-1)$ possible LOCs given the reference natural hazard

Gate type	Representation and quantification	Description
a	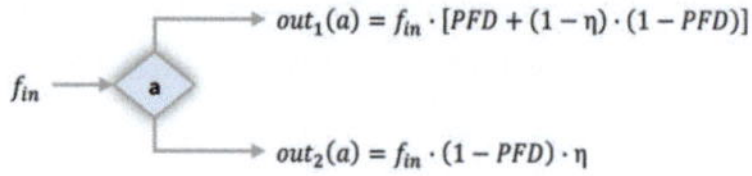	***Simple composite probability gate***. Unavailability, expressed as *PFD*, is combined with a single probability value for η
b	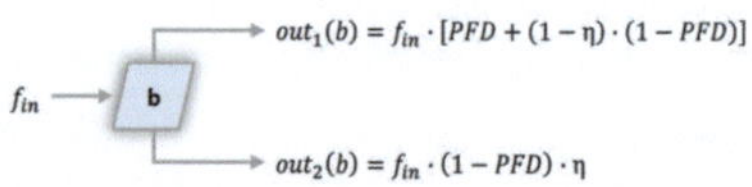	***Composite probability distribution gate***. Unavailability, expressed as *PFD*, is combined with a probability distribution expressing η. It is also possible to use an integrated value for η
c	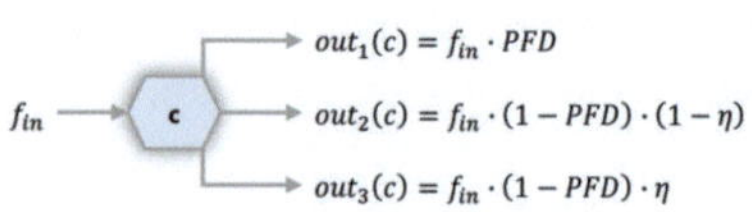	***Discrete probability distribution gate*** Depending on barrier, three or more events may originate
d	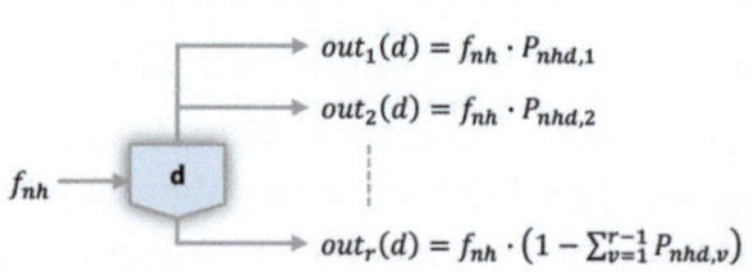	***Vessel fragility gate*** Based on the resistance of the equipment to the reference natural hazard, a set of $(r-1)$ failure outcomes can be originated, each with a related probability P_{nhd}, plus the r-th outcome representing the safe state

Adapted from Misuri et al. (2023)

Table 3.8 Reference ignition probabilities of flammable liquids and gases in accident scenarios triggered by earthquakes, floods, and lightning strikes

Natural hazard	Ignition probability (%)	
	Flammable liquids	Flammable gases
Earthquake	5.8	32.1
Flood	1.3	4.1
Lightning	49.8	52.8

Adapted from Ricci et al. (2021)

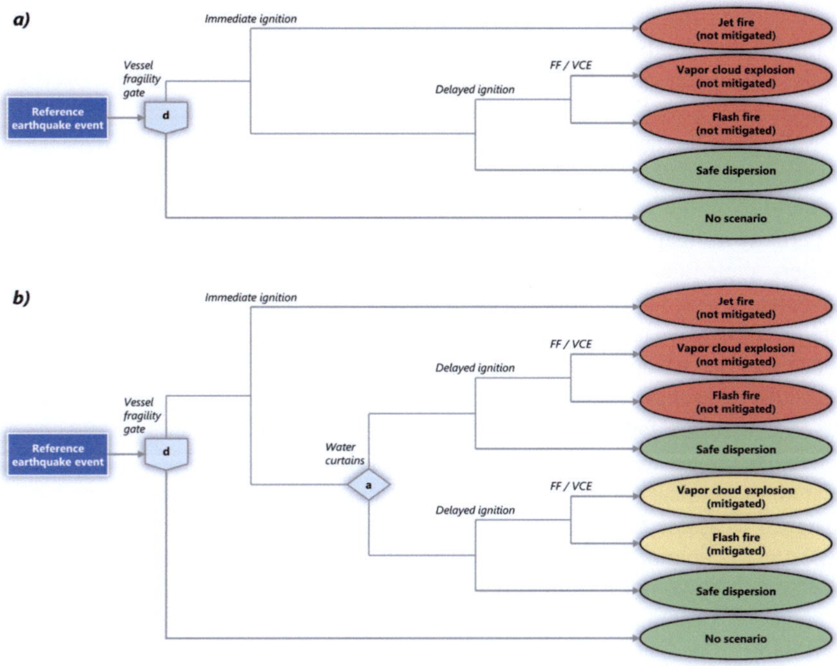

Fig. 3.5 Qualitative event trees reporting the primary Natech scenarios involving a pressurized vessel exposed to an earthquake event considering the continuous leak as release state. **a** No safety barriers are considered. **b** Water curtains are considered as safety systems acting on the tank. Adapted from Misuri et al. (2023)

continuous leak considered as the release state. In the top figure (panel a), the analysis does not incorporate any safety barriers for the tank. Conversely, in the bottom figure (panel b), the presence of water curtains as a safety system is taken into account, resulting in the identification of two additional partially mitigated scenarios.

3.2.6 Consequence Analysis of Credible Scenarios

Following the frequency calculations, a consequence analysis needs to be carried out (Step 6 in Fig. 3.1). This uses consolidated models from the literature (Center for Chemical Process Safety 2000; Mannan 2005; Bosh and Weterings 2005) to evaluate the physical effects of the end-point events considered in the technological scenario of concern (e.g., heat radiation from fires, concentrations of flammable or toxic substances during dispersions, and overpressure during explosions). The unique circumstances taking place in some Natech scenarios, such as flooding or strong winds, might necessitate a tailored analysis and specific assumptions, different from those used in the traditional Quantitative Risk Assessment (QRA). For instance, in the case of floods catch basins may lose their effectiveness in containing liquid pools. In dispersion models, selecting roughness lengths typical of water surfaces (Chamberlain 1983) is crucial to consider the presence of floodwaters.

Moreover, the analysis should also assess the impact of safety barriers on the consequences of accident scenarios. The "What if?" methodology not only supports scenario identification, but also the understanding of the outcomes of unmitigated or partially mitigated scenarios, taking into account the array of implemented safety barriers, their functions, and their operational state during an event. For example, as shown in Fig. 3.5, the application of What-if may support the assessment of the effectiveness of water curtains in reducing the concentration of flammable vapors. Successful barrier operation leads to considering a reduction in the concentrations dispersed in air downwind of the curtain, whereas no reduction should take place in the case of barrier failure.

Human targets are typically considered the primary risk receptors in QRA studies related to chemical and process industries. Consequently, vulnerability models are designed to assess the likelihood of damage to humans (usually death) following exposure to a dose of a specific physical effect. These models estimate the probability that fatalities take place using probit functions and threshold-based methodologies (Center for Chemical Process Safety 2000; Finney 1971; Van Den Bosh 1992), starting from the physical effects related to the technological scenarios of concern. The human vulnerability models used in traditional QRA can also be applied in Natech QRA scenarios.

3.2.7 *Identification of Credible Combinations of Events*

The possible development of multiple simultaneous outcomes is a distinctive aspect of Natech accidents (Cozzani and Salzano 2017). Integrating this aspect into Natech QRA methodology is essential because certain natural hazards, such as earthquakes and floods, can affect a significant number of industrial assets simultaneously. It is worth noting that not all natural hazards may lead to multiple failures. For instance, although multiple lightning strikes may occur during a storm, Natech QRA typically assumes that each strike affects only a single piece of equipment, based on past accident analyses (Necci et al. 2016).

The methodology for identifying and characterizing credible combinations of events draws from established methods used in domino effect QRA (Cozzani et al. 2005; Antonioni et al. 2009b). The approach involves assuming that the probability of damage to each critical target identified in Step 2 is independent of the failure of any other elements included in the analysis (Cozzani et al. 2014; Antonioni et al. 2007, 2015).

Considering the presence of n critical items, a single Natech scenario can involve the contemporary damage of l out of n units resulting in l final outcomes (more specifically, a final outcome per each item involved). Thus, considering that m_l

possible outcomes are identified by the ETA for each generic l-th target impacted, the total number of hazardous combinations N_{Natech} is assessed by Eq. (3.1):

$$N_{Natech} = \prod_{l=1}^{n} m_l - 1 \tag{3.1}$$

The number of combinations calculated through Eq. (3.1) does not consider the scenario where all critical equipment items are in a safe state, since it does not pose any risk from a human safety standpoint.

3.2.8 Frequency Evaluation of Combinations of Events

Once all credible combinations of events are identified, the procedure requires the calculation of the frequency of each combination of events. Every Natech scenario can be depicted by a vector $\boldsymbol{N^n}$ consisting of n elements, signifying the combination of events involving each of the n potential targets during the reference natural hazard. Consequently, as per Eq. (3.1), there are N_{Natech} distinct $\boldsymbol{N^n}$ vectors possible during the reference natural event. Each N_l^n element of a vector $\boldsymbol{N^n}$ denotes the condition of the l-th target in the considered Natech scenario. Consequently, the overall probability of a typical Natech scenario $P(\boldsymbol{N^n})$ can be determined using Eq. (3.2):

$$P\left(\boldsymbol{N^n}\right) = \prod_{l=1}^{n} P\left(N_l^n\right) \tag{3.2}$$

where $P\left(N_l^n\right)$ is the probability of a given state of the l-th target, assessed using the ETA of Sect. 3.2.5. Noteworthy, the gates outlined in Fig. 3.4 and Table 3.7 lead to the determination of the frequency of each state for a particular l-th target. To derive a probability value $P\left(N_l^n\right)$ for a specific state of the l-th target for use in Eq. (3.2), a unit value of f_{nh} should be applied in gate "d".

The actual frequency of the natural hazard f_{nh}, obtained in Step 1 (see Sect. 3.2.1), is introduced to compute the frequency of a typical overall Natech scenario $f(\boldsymbol{N^n})$ following Eq. (3.3):

$$f\left(\boldsymbol{N^n}\right) = f_{nh} \cdot P\left(\boldsymbol{N^n}\right) \tag{3.3}$$

3.2.9 Calculation of Overall Consequences and Vulnerability

The consequences of the overall Natech scenarios identified in Step 7 (see Sect. 3.2.7) should be assessed taking into account that combinations of multiple simultaneous heterogeneous outcomes are possible. Conventional consequence assessment models are not directly applicable to combinations of events. To address this challenge, the procedure developed for domino effects was extended to Natech QRA (Cozzani et al. 2014; Antonioni et al. 2009a, 2015; Cozzani et al. 2005). The approach involves estimating the consequences of the combination of events in each point as the sum, with an upper limit of 1, of the probability of human fatalities resulting from each outcome involved in the $\boldsymbol{N^n}$ Natech scenario. Thus, the death probability $V\left(\boldsymbol{N^n}\right)$ in a generic position related to the $\boldsymbol{N^n}$ Natech scenario is computed as the sum of the death probabilities related to each of the outcomes taking part in it:

$$V\left(\boldsymbol{N^n}\right) = \min\left[\sum_{l=1}^{n} V\left(N_l^n\right), 1\right] \tag{3.4}$$

where $V\left(N_l^n\right)$ is the death probability related to the state considered in the $\boldsymbol{N^n}$ Natech scenario of the l-th target.

3.2.10 Calculation of Risk Indexes

The assessment of the overall risk can be carried out once the overall Natech scenarios are characterized. Consistent with conventional approaches to chemical and process QRA, the risk level is typically expressed in terms of individual risk and societal risk (Center for Chemical Process Safety 2000; Uijt de Haag and Ale 2005; Mannan 2005). The former can be quantified by mapping local-specific individual risk (LSIR) using standardized procedures, while the latter can be represented by F/N plots, where $F(N)$ denotes the cumulative frequency of scenarios causing N or more expected fatalities, derived directly from the frequency $f(N)$ of scenarios causing N fatalities (Center for Chemical Process Safety 2000; Uijt de Haag and Ale 2005; Mannan 2005). To provide a concise representation of societal risk metrics, particularly beneficial for decision-making processes, two additional risk indices can be evaluated (Hirst and Carter 2002; Carter and Hirst 2000): the Potential Loss of Life (*PLL*) and Expected Value (*EV*), calculated respectively through Eqs. (3.5) and (3.6):

$$PLL = \sum_{i=1}^{m} f_i\left(N_i\right) \cdot N_i \tag{3.5}$$

$$EV = \sum_{i=1}^{m} f_i\left(N_i\right) \cdot N_i^{\alpha} \text{ with } \alpha > 1 \tag{3.6}$$

where i represent the technological scenario of concern. The *PLL* expresses the annual frequency of a fatality occurring within the area of interest. The *EV*, instead, is a composite index, attributing a higher risk value to severe accidents leading to a significant number of fatalities within the area of interest. The standardized procedures and tools commonly used within the framework of conventional QRA can be employed to evaluate the risk metrics (Cozzani et al. 2014, 2005; Antonioni et al. 2009b; Egidi et al. 1995; Spadoni et al. 2000).

3.3 Domino Effects in Natech Scenarios

Natech events are sequences of accidents initiated by natural hazards. The overall sequence of events is visually represented in Fig. 3.6. The root causes of the accidents directly stem from natural hazards, leading to technological scenarios (depicted by the red arrow in Fig. 3.6). These accidents are linked to effects on the site of natural hazards, such as seismic or wind loads. The accident sequence may halt after the primary technological scenario is triggered (green arrow in Fig. 3.6), or it may escalate due to domino effects (yellow arrow in Fig. 3.6), as observed in several Natech events that occurred in the past (Krausmann and Cruz 2013a; Cosmo Oil 2011).

Several methodologies exist in the literature for the quantitative risk assessment of domino effects in Natech events. Misuri et al. (2020a) developed a method for assessing the domino effect in Natech scenarios triggered by lightning, while Zeng et al. (2021) and Yang et al. (2020) concentrated on flood hazards. Ricci et al. (2024a) devised a procedure for assessing the risk of primary Natech and escalation scenarios triggered by interface fires and wildfires. Most of the procedures stemmed from those developed for domino effects in conventional accidents (Cozzani et al. 2014).

However, the likelihood of domino effects during Natech events is greatly influenced by the impact of natural hazards on safety barriers and systems in place to mitigate accidents. Actually, the impact of natural hazards on safety barriers and

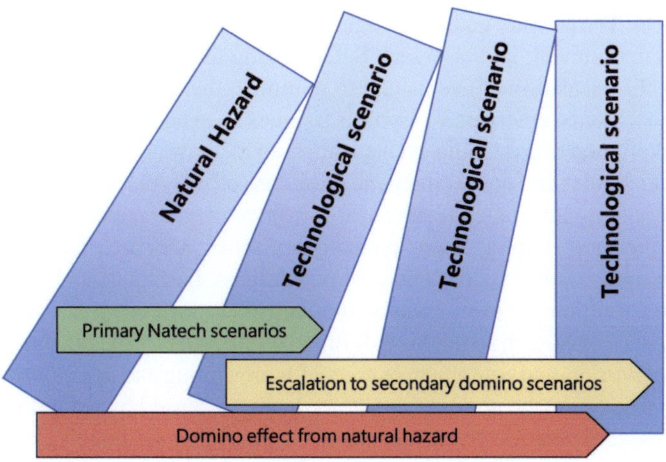

Fig. 3.6 Conceptualization of domino effect induced by natural hazards. Adapted from Misuri and Cozzani (2021a)

safety systems may hinder or deplete the prevention or mitigation actions in place to prevent domino effects during conventional scenarios. In the case safety barriers are compromised by the natural event, an increase in both frequency and consequence of technological escalation scenarios is expected (Misuri et al. 2023, 2020b; Misuri and Cozzani 2021b), underscoring the importance of evaluating domino effects in Natech accidents. The inclusion of safety barrier performance in domino effects assessment is therefore crucial to provide the actual risk figures considering the possible escalation of the primary technological scenario in Natech events.

Therefore, in recent years a specific effort was dedicated to the development of approaches aimed at considering the possible safety barrier degradation due to the action of the natural hazards in Natech scenarios (Misuri et al. 2021b, c). Specific methodologies were developed to include the effect of depleted safety barriers in the QRA of Natech events, both considering the possibility of unmitigated primary events and of domino effects (Misuri et al. 2023, 2021b, c). For instance, a method was developed by Misuri et al. (2021c) aimed at evaluating the frequency modification of domino effects due to the reduced performance of active and passive safety barriers. The same authors also proposed a method for assessing the risk increase due to the specific degradation of safety barriers in Natech scenarios Misuri et al. (2021b). It is important to remark that a specific assessment of the performance of safety barriers during Natech events is required to obtain representative risk figures (Misuri et al. 2020b).

3.4 Emergency Response in Natech Scenarios

Emergency response plays a crucial role in the framework of risk assessment and management. The primary goal of emergency teams relates to the protection of human lives, the environment, and/or assets. Safeguarding assets is particularly important to prevent the escalation of events and to minimize the consequences of accidents. In Natech accidents, emergency response assumes an even higher importance due to the high likelihood of cascading events compared to conventional scenarios. Natural disasters can hinder and delay emergency response efforts, placing strain on systems that must concurrently contend with hazards from both natural and technological disasters. Furthermore, natural events typically impact broader areas beyond industrial sites, encompassing nearby regions with populations, buildings, and infrastructure, thereby constraining the availability and effectiveness of emergency teams. Emergency response in Natech accidents is further complicated by the potential depletion of technical barriers and equipment essential for preventing and mitigating accidental scenarios, which may be unavailable due to damages caused by natural events.

Currently, the literature offers only a limited number of methodologies for assessing emergency response, with many focusing on specific aspects. For instance, Bernier et al. (2019) devised a method to evaluate site accessibility during storm surge events. Baser and Behnam (2020) introduced a methodology to assess the impact of emergency response on the failure probability of equipment items in the

Table 3.9 Outline of the methodology proposed by Ricci et al. (2024b) for the assessment of emergency response performance and time in cascading sequences triggered by natural events

Step	Definition	Description
Step 0	Definition of the phases of emergency response	Schematization of the emergency response in a sequence of consecutive actions necessary to effectively carry out to complete the procedure
Step 1	Definition of emergency response factors	Identification of the relevant emergency response factors (i.e., action carried out by the emergency responders, equipment required to carry out the activities, technical needs that contribute to the emergency response)
Step 2	Characterization of baseline factors	Quantitative evaluation of the performance of the identified factors based on expert elicitation
Step 3	Assessment of the emergency response performance	Definition of the Bayesian Network model representing the emergency response and quantification of the model to derive performance values
Step 4	Assessment of the time required for emergency response	Quantification of the time required to carry out the emergency response procedure through Monte Carlo simulations

context of cascading post-earthquake fires. This approach addresses criticalities associated with emergency response, such as the challenge of extinguishing multiple fires simultaneously. Ricci et al. (2024b) developed a methodology for the evaluation of emergency response performance and time in fire-induced cascading scenarios initiated by natural events. Table 3.9 provides a concise overview of the methodology, outlining and detailing its key steps. Further details on the application of the methodology are reported in the original study (Ricci et al. 2024b).

The methodology outlined in Table 3.9 can be applied to a particular industrial site of interest, provided that specific information on influencing factors is available and a tailored survey is conducted. In Ricci et al. (2024b), a comprehensive set of relevant factors is proposed as a baseline list, and an expert elicitation was conducted to gather a general assessment of emergency response and baseline quantitative data regarding the success probability and the required time for response in the event of earthquakes and floods. Table 3.10 presents the % performance of emergency

Table 3.10 Baseline results for the performance of the different phases and emergency response in the case of natural events with respect to that estimated for conventional accidents (Ricci et al. 2024b). Performance$_e$: % performance in the case of cascading sequences triggered by earthquakes. Performance$_f$: % performance in the case of cascading sequences triggered by floods

Phase	Definition	$Performance_e$ (%)	$Performance_f$ (%)
Detection	Begins when a fire develops and ends when its detection is communicated to the control room	59.9	73.3
Notification	Starts when the control room is aware of the fire and ends when the alarm is notified to the emergency teams	49.9	64.0
Turnout	Begins when the accident is notified to the emergency teams and ends when the teams leave the fire station	53.3	70.0
Routing	Starts when the emergency teams leave the fire station and ends when they reach the scene	33.7	39.4
Setup	Begins once the emergency teams arrive at the accident scene and ends when they start fighting the fire	51.0	53.9
Fighting	Starts when emergency teams begin to counteract the fire (firefighting, protecting neighboring tanks, etc.) and ends when the fire is completely extinguished	39.5	50.6
Overall emergency response	**From detection to fighting phases**	**2.6**	**6.6**

response phases in scenarios involving earthquakes and floods with respect to that estimated in conventional accidents.

The model developed by Ricci et al. (2024b) serves as a valuable tool for enhancing the emergency management of Natech scenarios, in particular in the context

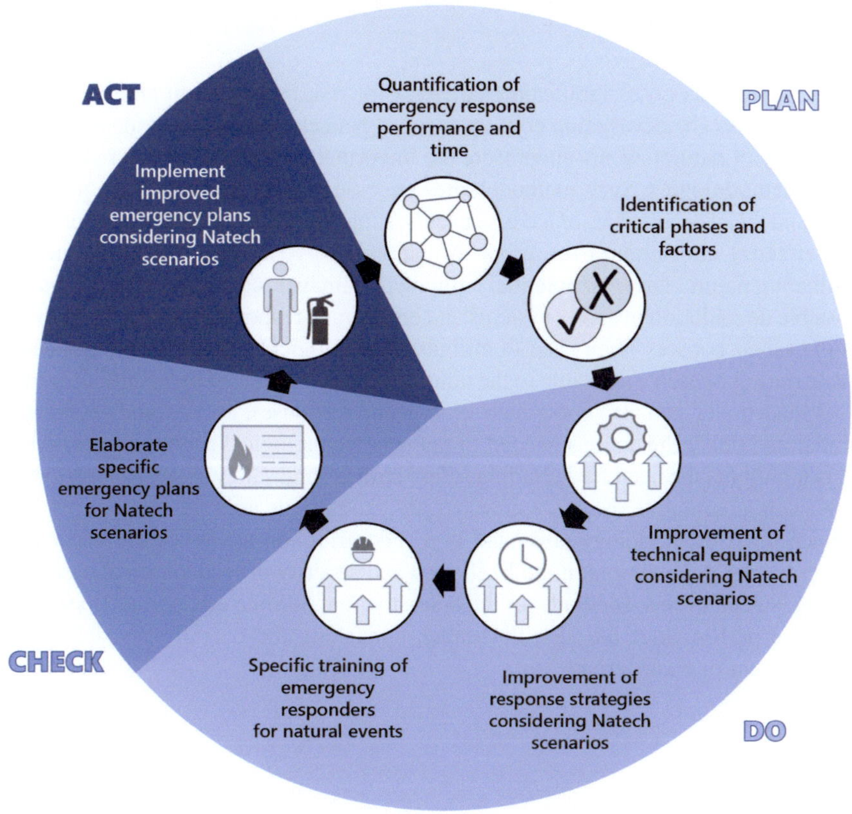

Fig. 3.7 Example of Plan-Do-Check-Act cycle aimed at improving performance and time in emergency response in cascading events caused by natural events. Adapted from Ricci et al. (2024b)

of emergency planning. Figure 3.7 illustrates a Plan-Do-Check-Act (PDCA) cycle designed to improve the specific performance of emergency response in Natech scenarios. Beginning with the assessment of the current status of emergency response, the methodology enables the identification of the phases and factors of emergency response most affected by natural events. Thus, the more critical technical features and strategies can be enhanced, and responder training may be improved. Subsequently, specific emergency plans for Natech scenarios can be developed and implemented. Once these changes are implemented, the overall performance of emergency response can be reassessed, providing updated figures related to the probability of success and the time of emergency response. This iterative cycle can continue until a target value for the emergency response time is achieved.

3.5 Conclusions

Natech accidents pose significant challenges for risk management and emergency preparedness. The devastating consequences of Natech accidents and the increasing frequency of natural events underscore the importance of robust Natech risk assessment methodologies. Such methodologies are essential for evaluating the potential risks and the consequences of adverse events, supporting informed decision-making, and effective allocation of resources. A comprehensive assessment should include evaluating natural hazards, potential equipment damages, cascading effects, and the possible degradation of safety systems and emergency response. Indeed, the conventional safety barriers often used to mitigate technological accidents may be inadequate during Natech events due to the unique conditions created by natural disasters. This inadequacy can also complicate emergency responses, as natural disasters can hinder and delay the actions needed to manage technological accidents effectively. This highlights the need for including safety barriers and safety systems degradation in Natech quantitative risk assessments (QRA) to provide accurate risk figures. The Natech QRA methodology presented in this contribution addresses the impacts of safety barrier degradation on the likelihood and severity of technological scenarios during Natech events. Methods for the assessment of domino effects and emergency response performance are also illustrated, enhancing the overall approach to the management of Natech risks.

References

American Petroleum Institute. (2016a) API RP 580: Risk based inspection.

American Petroleum Institute. (2016b). API RP 581: Risk-based inspection technology.

Andersen, H., Casal, J., Dandrieux, A., Debray, B., De Dianous, V., Duijm, N. J., Delvosalle, C., Fievez, C., Goossens, L., Gowland, R. T., Hale, A. J., Hourtolou, D., Mazzarotta, B., Pipart, A., Planas, E., Prats, F., Salvi, O., & Tixier, J. (2004). *Aramis: Accidental risk assessment methodology for industries in the context of the Seveso II directive-User guide*. The European Commission - Energy, Environment and Sustainable Development: Technical report.

Antonioni, G., Spadoni, G., & Cozzani, V. (2007). A methodology for the quantitative risk assessment of major accidents triggered by seismic events. *Journal of Hazardous Materials, 147*(1), 48–59. https://doi.org/10.1016/j.jhazmat.2006.12.043

Antonioni, G., Spadoni, G., & Cozzani, V. (2009). Application of domino effect quantitative risk assessment to an extended industrial area. *Journal of Loss Prevention in the Process Industries, 22*(5), 614–624. https://doi.org/10.1016/j.jlp.2009.02.012. Quantitative Risk Analysis-Special issue devoted to Norberto Piccinini.

Antonioni, G., Bonvicini, S., Spadoni, G., & Cozzani, V. (2009). Development of a framework for the risk assessment of Na-tech accidental events. *Reliability Engineering & System Safety, 94*(9), 1442–1450. https://doi.org/10.1016/j.ress.2009.02.026

Antonioni, G., Landucci, G., Necci, A., Gheorghiu, D., & Cozzani, V. (2015). Quantitative assessment of risk due to Natech scenarios caused by floods. *Reliability Engineering & System Safety, 142*, 334–345. https://doi.org/10.1016/j.ress.2015.05.020

Aureli, F., Dazzi, S., Maranzoni, A., Mignosa, P., & Vacondio, R. (2015). Experimental and numerical evaluation of the force due to the impact of a dam-break wave on a structure. *Advances in Water Resources, 76*, 29–42. https://doi.org/10.1016/j.advwatres.2014.11.009

Basco, A., & Salzano, E. (2017). The vulnerability of industrial equipment to tsunami. *Journal of Loss Prevention in the Process Industries, 50*, 301–307. https://doi.org/10.1016/j.jlp.2016.11.009

Baser, B., & Behnam, B. (2020). An emergency response plan for cascading post-earthquake fires in fuel storage facilities. *Journal of Loss Prevention in the Process Industries, 65*, 104155. https://doi.org/10.1016/j.jlp.2020.104155

Bernier, C., Gidaris, I., Balomenos, G. P., & Padgett, J. E. (2019). Assessing the accessibility of petrochemical facilities during storm surge events. *Reliability Engineering & System Safety, 188*, 155–167. https://doi.org/10.1016/j.ress.2019.03.021

Cadwallader, L. C. (1996). Fire protection system operating experience review for fusion applications. *Fusion Technology, 30*(3P2B), 1420–1423. https://doi.org/10.13182/FST96-A11963147

Campedel, M., Cozzani, V., Garcia-Agreda, A., & Salzano, E. (2008). Extending the quantitative assessment of industrial risks to earthquake effects. *Risk Analysis, 28*(5), 1231–1246. https://doi.org/10.1111/j.1539-6924.2008.01092.x

Caratozzolo, V., Misuri, A., & Cozzani, V. (2022). A generalized equipment vulnerability model for the quantitative risk assessment of horizontal vessels involved in Natech scenarios triggered by floods. *Reliability Engineering & System Safety, 223*, 108504. https://doi.org/10.1016/j.ress.2022.108504

Carter, D. A., & Hirst, I. L. (2000). 'Worst case' methodology for the initial assessment of societal risk from proposed major accident installations. *Journal of Hazardous Materials, 71*(1), 117–128. https://doi.org/10.1016/S0304-3894(99)00075-8

Cecil, D. J., Buechler, D. E., & Blakeslee, R. J. (2014). Gridded lightning climatology from TRMM-LIS and OTD: Dataset description. *Atmospheric Research, 135–136*, 404–414. https://doi.org/10.1016/j.atmosres.2012.06.028

Center for Chemical Process Safety. (2000). *Guidelines for chemical process quantitative risk analysis*. American Institute of Chemical Engineers - Center of Chemical Process Safety.

Center for Chemical Process Safety. (2001). *Layer of protection analysis—Simplified process risk assessment*. American Institute of Chemical Engineers—Center of Chemical Process Safety.

Centre for Research on the Epidemiology of Disasters. (2020) *EM-DAT: The emergency events database*. Université Catholique de Louvain. www.emdat.be

Centro Internacional de Métodos Numéricos en Ingeniería (CIMNE), Ingenieros - Arquitectos, Consultores (INGENIAR), Ingeniería Técnica y Científica (ITEC), and Engineering Advise International (EAI). (2013). *Probabilistic modelling of natural risks at the global level: Global risk model*. Technical report, The United Nations Office for Disaster Risk Reduction - Global Assessment Report on Disaster Risk Reduction.

Chamberlain, A. C. (1983). Roughness length of sea, sand, and snow. *Boundary-Layer Meteorology, 25*(4), 405–409. https://doi.org/10.1007/BF02041157

Cosmo Oil. (2011). Overview of the fire and explosion at Chiba refinery, the cause of the accident and the action plan to prevent recurrence. https://ceh.cosmo-oil.co.jp/eng/press/110802/index.html

Cozzani, V., & Salzano, E. (2017). Chapter 6—Technological hazard characterization. In E. Krausmann, Cruz, A. M., & E. Salzano (Eds.), *Natech risk assessment and management* (pp. 91-103). Elsevier. https://doi.org/10.1016/B978-0-12-803807-9.00006-1

Cozzani, V., Gubinelli, G., Antonioni, G., Spadoni, G., & Zanelli, S. (2005). The assessment of risk caused by domino effect in quantitative area risk analysis. *Journal of Hazardous Materials, 127*(1), 14–30. https://doi.org/10.1016/j.jhazmat.2005.07.003

Cozzani, V., Campedel, M., Renni, E., & Krausmann, E. (2010). Industrial accidents triggered by flood events: Analysis of past accidents. *Journal of Hazardous Materials, 175*(1), 501–509. https://doi.org/10.1016/j.jhazmat.2009.10.033

Cozzani, V., Antonioni, G., Landucci, G., Tugnoli, A., Bonvicini, S., & Spadoni, G. (2014). Quantitative assessment of domino and Natech scenarios in complex industrial areas. *Journal of Loss Prevention in the Process Industries, 28*, 10–22. https://doi.org/10.1016/j.jlp.2013.07.009

Cruz, A. M., & Steinberg, L. J. (2005). Industry preparedness for earthquakes and earthquake-triggered hazmat accidents in the 1999 Kocaeli earthquake. *Earthquake Spectra, 21*(2), 285–303. https://doi.org/10.1193/1.1889442

de Moel, H., van Alphen, J., & Aerts, J. (2009). Flood maps in Europe-methods, availability and use. *Natural Hazards and Earth System Sciences, 9*(2), 289–301. https://doi.org/10.5194/nhess-9-289-2009

Egidi, D., Foraboschi, F. P., Spadoni, G., & Amendola, A. (1995). The Aripar project: Analysis of the major accident risks connected with industrial and transportation activities in the Ravenna area. *Reliability Engineering & System Safety, 49*(1), 75–89. https://doi.org/10.1016/0951-8320(95)00026-X

Fabbrocino, G., Iervolino, I., Orlando, F., & Salzano, E. (2005). Quantitative risk analysis of oil storage facilities in seismic areas. *Journal of Hazardous Materials, 123*(1), 61–69. https://doi.org/10.1016/j.jhazmat.2005.04.015

Finney, D. J. (1971). *Probit analysis* (3rd ed.). Cambridge University Press.

Finney, M. A. (1998). *FARSITE: Fire area simulator-model development and evaluation*. U.S: Department of Agriculture, Forest Service, Rocky Mountain Research Station. https://doi.org/10.2737/rmrs-rp-4

Frank, K., Gravestock, N., Spearpoint, M., & Fleischmann, C. (2013). A review of sprinkler system effectiveness studies. *Fire Science Reviews, 2*(1), 6. https://doi.org/10.1186/2193-0414-2-6

Girgin, S. (2011). The Natech events during the 17 August 1999 Kocaeli earthquake: Aftermath and lessons learned. *Natural Hazards and Earth System Sciences, 11*(4), 1129–1140. https://doi.org/10.5194/nhess-11-1129-2011

Global Facility for Disaster Reduction and Recovery. (2024). *ThinkHazard!*. https://thinkhazard.org/en/

HAZUS. (1997). *Earthquake loss estimation methodology*. Risk Management Solutions: National Institute of Building Science.

Hilton, J., Hetherton, L., Miller, C., Sullivan, A., & Prakash, M. (2014). The spark framework. Technical report, Report Number EP152898. CSIRO Digital Productivity Flagship, Victoria, Australia.

Hirst, I. L., & Carter, D. A. (2002). A "worst case" methodology for obtaining a rough but rapid indication of the societal risk from a major accident hazard installation. *Journal of Hazardous Materials, 92*(3), 223–237. https://doi.org/10.1016/S0304-3894(02)00016-X

Iervolino, I., Giorgio, M., & Polidoro, B. (2014). Sequence-based probabilistic seismic hazard analysis. *Bulletin of the Seismological Society of America, 104*(2), 1006–1012. https://doi.org/10.1785/0120130207

Iervolino, I., Chioccarelli, E., Giorgio, M., Marzocchi, W., Zuccaro, G., Dolce, M., & Manfredi, G. (2015). Operational (short-term) earthquake loss forecasting in Italy. *Bulletin of the Seismological Society of America, 105*(4), 2286–2298. https://doi.org/10.1785/0120140344

Iervolino, I., Giorgio, M., & Chioccarelli, E. (2016). Markovian modeling of seismic damage accumulation. *Earthquake Engineering & Structural Dynamics, 45*(3), 441–461. https://doi.org/10.1002/eqe.2668

International Electrotechnical Commission. (2006). *Protection against lightning-Part 2: Risk management - 62305*. International Electrotechnical Commission: Technical report.

International Electrotechnical Commission. (2019). *Lightning density based on lightning location systems (LLS)-General principles-62858*. International Electrotechnical Commission: Technical report.

International Electrotechnical Commission. (2010). IEC 61508: Functional safety of electrical/electronic/programmable electronic safety-related systems.

International Electrotechnical Commission. (2016). IEC 61511: Functional safety—Safety instrumented systems for the process industry sector.

Kameshwar, S., & Padgett, J. E. (2015). Fragility assessment of above ground petroleum storage tanks under storm surge. In *12th International Conference on Applications of Statistics and Probability in Civil Engineering*.

Kameshwar, S., & Padgett, J. E. (2018). Storm surge fragility assessment of above ground storage tanks. *Structural Safety, 70*, 48–58. https://doi.org/10.1016/j.strusafe.2017.10.002

Khakzad, N., & Van Gelder, P. (2017). Fragility assessment of chemical storage tanks subject to floods. *Process Safety and Environmental Protection, 111*, 75–84. https://doi.org/10.1016/j.psep.2017.06.012

Khakzad, N., Van Gelder, P. (2018). Vulnerability of industrial plants to flood-induced natechs: A bayesian network approach. *Reliability Engineering & System Safety, 169*, 403–411. https://doi.org/10.1016/j.ress.2017.09.016

Elisabeth Krausmann and Ana Maria Cruz. (2013). Impact of the 11 March 2011, Great East Japan earthquake and tsunami on the chemical industry. *Natural Hazards, 67*(2), 811–828. https://doi.org/10.1007/s11069-013-0607-0

Krausmann, E., & Salzano, E. (2017). Chapter 3—Lessons learned from Natech events. In E. Krausmann, A. M. Cruz, & E. Salzano (Eds.), *Natech risk assessment and management* (pp. 33–52). Elsevier. https://doi.org/10.1016/B978-0-12-803807-9.00003-6

Krausmann, E., Cruz, A. M., & Affeltranger, B. (2010). The impact of the 12 May 2008 Wenchuan earthquake on industrial facilities. *Journal of Loss Prevention in the Process Industries, 23*(2), 242–248. https://doi.org/10.1016/j.jlp.2009.10.004

Krausmann, E., Renni, E., Campedel, M., & Cozzani, V. (2011). Industrial accidents triggered by earthquakes, floods and lightning: Lessons learned from a database analysis. *Natural Hazards, 59*(1), 285–300. https://doi.org/10.1007/s11069-011-9754-3

Landucci, G., Molag, M., Reinders, J., & Cozzani, V. (2009). Experimental and analytical investigation of thermal coating effectiveness for 3m3 LPG tanks engulfed by fire. *Journal of Hazardous Materials, 161*(2), 1182–1192. https://doi.org/10.1016/j.jhazmat.2008.04.097

Landucci, G., Antonioni, G., Tugnoli, A., & Cozzani, V. (2012). Release of hazardous substances in flood events: Damage model for atmospheric storage tanks. *Reliability Engineering & System Safety, 106*, 200–216. https://doi.org/10.1016/j.ress.2012.05.010

Landucci, G., Necci, A., Antonioni, G., Tugnoli, A., & Cozzani, V. (2014). Release of hazardous substances in flood events: Damage model for horizontal cylindrical vessels. *Reliability Engineering & System Safety, 132*, 125–145. https://doi.org/10.1016/j.ress.2014.07.016

Landucci, G., Argenti, F., Tugnoli, A., & Cozzani, V. (2015). Quantitative assessment of safety barrier performance in the prevention of domino scenarios triggered by fire. *Reliability Engineering & System Safety, 143*, 30–43. https://doi.org/10.1016/j.ress.2015.03.023

Landucci, G., Argenti, F., Spadoni, G., & Cozzani, V. (2016). Domino effect frequency assessment: The role of safety barriers. *Journal of Loss Prevention in the Process Industries, 44*, 706–717. https://doi.org/10.1016/j.jlp.2016.03.006

Landucci, G., Necci, A., Antonioni, G., Argenti, F., & Cozzani, V. (2017). Risk assessment of mitigated domino scenarios in process facilities. *Reliability Engineering & System Safety, 160*, 37–53. https://doi.org/10.1016/j.ress.2016.11.023

Lanzano, G., Salzano, E., de Magistris, F. S., & Fabbrocino, G. (2013). Seismic vulnerability of natural gas pipelines. *Reliability Engineering & System Safety, 117*, 73–80. https://doi.org/10.1016/j.ress.2013.03.019

Lanzano, G., Salzano, E., de Magistris, F. S., & Fabbrocino, G. (2014). Seismic vulnerability of gas and liquid buried pipelines. *Journal of Loss Prevention in the Process Industries, 28*, 72–78. https://doi.org/10.1016/j.jlp.2013.03.010

Lindell, M. K., & Perry, R. W. (1996). Identifying and managing conjoint threats: Earthquake-induced hazardous materials releases in the US. *Journal of Hazardous Materials, 50*(1), 31–46. https://doi.org/10.1016/0304-3894(96)01764-5

Lorito, S., Selva, J., Basili, R., Romano, F., Tiberti, M. M., & Piatanesi, A. (2014). Probabilistic hazard for seismically induced tsunamis: Accuracy and feasibility of inundation maps. *Geophysical Journal International, 200*(1), 574–588. https://doi.org/10.1093/gji/ggu408

Madonna, M., Martella, G., Monica, L., Maini, E. P., & Tomassini, L. (2009). The human factor in risk assessment: Methodological comparison between human reliability analysis techniques. *Prevention Today, 5*, 67–83.

Mannan, S. (2005). *Lees' loss prevention in the process industries*. Elsevier Butterworth-Heinemann.

Mayorga, S. Z., Sánchez-Silva, M., Ramírez Olivar, O. J., & Giraldo, F. M. (2019). Development of parametric fragility curves for storage tanks: A natech approach. *Reliability Engineering & System Safety, 189*, 1–10. https://doi.org/10.1016/j.ress.2019.04.008

Mesa-Gómez, A., Casal, J., Sánchez-Silva, M., & Muñoz, F. (2021). Advances and gaps in Natech quantitative risk analysis. *Processes, 9*(1). https://doi.org/10.3390/pr9010040

Mignan, A., Wiemer, S., & Giardini, D. (2014). The quantification of low-probability-high-consequences events: Part I. *A generic multi-risk approach. Natural Hazards, 73*(3), 1999–2022. https://doi.org/10.1007/s11069-014-1178-4

Misuri, A., & Cozzani, V. (2021a). Chapter Six—Natural events driven domino effect. In F. Khan, V. Cozzani, & G. Reniers (Eds.), *Domino effect: Its prediction and prevention*, volume 5 of Methods in Chemical Process Safety (pp. 183–225). Elsevier. https://doi.org/10.1016/bs.mcps.2021.05.006

Misuri, A., Casson Moreno, V., Quddus, N., & Cozzani, V. (2019). Lessons learnt from the impact of hurricane harvey on the chemical and process industry. *Reliability Engineering & System Safety, 190*, 106521. ISSN 0951-8320. https://doi.org/10.1016/j.ress.2019.106521

Misuri, A., & Cozzani, V. (2021). A paradigm shift in the assessment of Natech scenarios in chemical and process facilities. *Process Safety and Environmental Protection, 152*, 338–351. https://doi.org/10.1016/j.psep.2021.06.018

Misuri, A., Antonioni, G., & Cozzani, V. (2020). Quantitative risk assessment of domino effect in Natech scenarios triggered by lightning. *Journal of Loss Prevention in the Process Industries, 64*, 104095. https://doi.org/10.1016/j.jlp.2020.104095

Misuri, A., Landucci, G., & Cozzani, V. (2020). Assessment of safety barrier performance in Natech scenarios. *Reliability Engineering & System Safety, 193*, 106597. https://doi.org/10.1016/j.ress.2019.106597

Misuri, A., Landucci, G., & Cozzani, V. (2021). Assessment of risk modification due to safety barrier performance degradation in Natech events. *Reliability Engineering & System Safety, 212*, 107634. https://doi.org/10.1016/j.ress.2021.107634

Misuri, A., Landucci, G., & Cozzani, V. (2021). Assessment of safety barrier performance in the mitigation of domino scenarios caused by Natech events. *Reliability Engineering & System Safety, 205*, 107278. https://doi.org/10.1016/j.ress.2020.107278

Misuri, A., Ricci, F., Sorichetti, R., & Cozzani, V. (2023). The effect of safety barrier degradation on the severity of primary Natech scenarios. *Reliability Engineering & System Safety, 235*, 109272. https://doi.org/10.1016/j.ress.2023.109272

Moschonas, I., Karakostas, C., Lekidis, V., & Papadopoulos, S. (2014). Investigation of seismic vulnerability of industrial pressure vessels. In *Second European Conference on Earhquake Engineering and Seismology*.

Munich, R. E. (2011). Nathan word map of natural hazards

Musson, R. (2000). Generalised seismic hazard maps for the Pannonian basin using probabilistic methods. *Pure and Applied Geophysics, 157*(1), 147–169. https://doi.org/10.1007/PL00001092

Naderpour, M., Mojaddadi, H. R., Khakzad, N., & Pradhan, B. (2019). Forest fire induced natech risk assessment: A survey of geospatial technologies. *Reliability Engineering & System Safety, 191*, 106558. https://doi.org/10.1016/j.ress.2019.106558

National Aeronautics and Space Administration. (2024). *EOSDIS worldview*. https://worldview.earthdata.nasa.gov/

Necci, A., Antonioni, G., Cozzani, V., Krausmann, E., Borghetti, A., & Alberto Nucci, C. (2013). A model for process equipment damage probability assessment due to lightning. *Reliability Engineering & System Safety, 115*, 91–99. https://doi.org/10.1016/j.ress.2013.02.018

Necci, A., Antonioni, G., Cozzani, V., Krausmann, E., Borghetti, A., & Alberto Nucci, C. (2014). Assessment of lightning impact frequency for process equipment. *Reliability Engineering & System Safety, 130*, 95–105. https://doi.org/10.1016/j.ress.2014.05.001

Necci, A., Antonioni, G., Bonvicini, S., & Cozzani, V. (2016). Quantitative assessment of risk due to major accidents triggered by lightning. *Reliability Engineering & System Safety, 154*, 60–72. https://doi.org/10.1016/j.ress.2016.05.009

New Zealand Fire Service Commission. (2008). *Effectiveness of fire safety systems for use in quantitative risk assessments-Research report number 89*. MARSH: Technical report.

Organisation for Economic Cooperation and Development. (2015). *Addendum number 2 to the OECD guiding principles for chemical accident prevention, preparedness and response (2nd ed.) to address natural hazards triggering technological accidents (Natechs)*. Technical report, Organisation for Economic Cooperation and Development.

O'Rourke, M. J., & So, P. (2000). Seismic fragility curves for on-grade steel tanks. *Earthquake Spectra, 16*(4), 801–815. https://doi.org/10.1193/1.1586140

Pamela Sands Showalter and Mary Fran Myers. (1994). Natural disasters in the United States as release agents of oil, chemicals, or radiological materials between 1980–1989: Analysis and recommendations. *Risk Analysis, 14*(2), 169–182. https://doi.org/10.1111/j.1539-6924.1994.tb00042.x

Pilotti, M., Tomirotti, M., Valerio, G., & Bacchi, B. (2010). Simplified method for the characterization of the hydrograph following a sudden partial dam break. *Journal of Hydraulic Engineering, 136*(10), 693–704. https://doi.org/10.1061/(ASCE)HY.1943-7900.0000231

Pilotti, M., Maranzoni, A., Tomirotti, M., & Valerio, G. (2011). 1923 Gleno dam break: Case study and numerical modeling. *Journal of Hydraulic Engineering, 137*(4), 480–492. https://doi.org/10.1061/(ASCE)HY.1943-7900.0000327

Qin, R., Zhu, J., & Khakzad, N. (2020). Multi-hazard failure assessment of atmospheric storage tanks during hurricanes. *Journal of Loss Prevention in the Process Industries, 68*, 104325. https://doi.org/10.1016/j.jlp.2020.104325

Ramírez Olivar, O. J., Mayorga, S. Z., Giraldo, F. M., Sánchez-Silva, M., Pinelli, J.-P., & Salzano, E. (2020). The effects of extreme winds on atmospheric storage tanks. *Reliability Engineering & System Safety, 195*, 106686. https://doi.org/10.1016/j.ress.2019.106686

Renni, E., Krausmann, E., & Cozzani, V. (2010). Industrial accidents triggered by lightning. *Journal of Hazardous Materials, 184*(1), 42–48. https://doi.org/10.1016/j.jhazmat.2010.07.118

Ricci, F., Moreno, V. C., & Cozzani, V. (2021). A comprehensive analysis of the occurrence of Natech events in the process industry. *Process Safety and Environmental Protection, 147*, 703–713. https://doi.org/10.1016/j.psep.2020.12.031

Ricci, F., Yang, M., Reniers, G., & Cozzani, V. (2022). The role of emergency response in risk management of cascading events caused by Natech accidents. *Chemical Engineering Transaction, 91*, 361–366. https://doi.org/10.3303/CET2291061

Ricci, F., Bonvicini, S., Moreno, V. C., & Cozzani, V. (2022). The quantitative assessment of damage to the environment in major accidents caused by natural events. *Chemical Engineering Transaction, 90*, 253–258. https://doi.org/10.3303/CET2290043

Ricci, F., Yang, M., Reniers, G., & Cozzani, V. (2024). Emergency response in cascading scenarios triggered by natural events. *Reliability Engineering & System Safety, 243*, 109820. https://doi.org/10.1016/j.ress.2023.109820

Ricci, F., Misuri, A., Scarponi, G. E., Cozzani, V., & Demichela, M. (2024). Vulnerability assessment of industrial sites to interface fires and wildfires. *Reliability Engineering & System Safety, 243*, 109895. https://doi.org/10.1016/j.ress.2023.109895

Roberts, T. A. (2004). Directed deluge system designs and determination of the effectiveness of the currently recommended minimum deluge rate for the protection of LPG tanks. *Journal of Loss Prevention in the Process Industries, 17*(2), 103–109. https://doi.org/10.1016/j.jlp.2003.10.005

Rossi, L., Moreno, V. C., & Landucci, G. (2022). Vulnerability assessment of process pipelines affected by flood events. *Reliability Engineering & System Safety, 219*, 108261. https://doi.org/10.1016/j.ress.2021.108261

Salzano, E., Iervolino, I., & Fabbrocino, G. (2003). Seismic risk of atmospheric storage tanks in the framework of quantitative risk analysis. *Journal of Loss Prevention in the Process Industries, 16*(5), 403–409. ISSN 0950-4230. https://doi.org/10.1016/S0950-4230(03)00052-4

Salzano, E., Agreda, A. G., Di Carluccio, A., & Fabbrocino, G. (2009). Risk assessment and early warning systems for industrial facilities in seismic zones. *Reliability Engineering & System Safety, 94*(10), 1577–1584. https://doi.org/10.1016/j.ress.2009.02.023

Salzano, E., Basco, A., Busini, V., Cozzani, V., Marzo, E., Rota, R., & Spadoni, G. (2013). Public awareness promoting new or emerging risks: Industrial accidents triggered by natural hazards (Natech). *Journal of Risk Research, 16*(3–4), 469–485. https://doi.org/10.1080/13669877.2012.729529

Selva, J. (2013). Long-term multi-risk assessment: statistical treatment of interaction among risks. *Natural Hazards, 67*(2), 701–722. https://doi.org/10.1007/s11069-013-0599-9

Selva, J., Tonini, R., Molinari, I., Tiberti, M. M., Romano, F., Grezio, A., Melini, D., Piatanesi, A., Basili, R., & Lorito, S. (2016). Quantification of source uncertainties in seismic probabilistic tsunami hazard analysis (SPTHA). *Geophysical Journal International, 205*(3), 1780–1803. https://doi.org/10.1093/gji/ggw107

Selva, J., Lorito, S., Volpe, M., Romano, F., Tonini, R., Perfetti, P., Bernardi, F., Taroni, M., Scala, A., Babeyko, A., Løvholt, F., Gibbons, S. J., Macías, J., Castro, M. J., González-Vida, J. M., Sánchez-Linares, C., Bayraktar, H. B., Basili, R., Maesano, F. E., & Amato, A. (2021). Probabilistic tsunami forecasting for early warning. *Nature Communications*. https://doi.org/10.1038/s41467-021-25815-w

Spadoni, G., Egidi, D., & Contini, S. (2000). Through Aripar-GIS the quantified area risk analysis supports land-use planning activities. *Journal of Hazardous Materials, 71*(1), 423–437. https://doi.org/10.1016/S0304-3894(99)00091-6

Steinberg, L. J., & Cruz, A. M. (2004). When natural and technological disasters collide: Lessons from the Turkey earthquake of August 17, 1999. *Natural Hazards Review, 5*(3), 121–130. https://doi.org/10.1061/(ASCE)1527-6988(2004)5:3(121)

Tymstra, C., Bryce, R. W., Wotton, B. M., Taylor, S. W., & Armitage, O. B. (2010). *Development and structure of Prometheus: The Canadian wildland fire growth simulation model*. Technical report, Northern Forestry Centre—Canadian Forest Service—Information Report NOR-X-417.

Uijt de Haag, P. A. M., & Ale, B. J. M. (2005). *Guidelines for quantitative risk assessment (Purple book)*. Committee for the Prevention of Disasters.

U.S. Chemical Safety and Hazard Investigation Board. (2018). *Organic peroxide decomposition, release, and fire at Arkema crosby following hurricane Harvey flooding. report number 2017-08-i-tx*. Technical report, U.S. Chemical Safety and Hazard Investigation Board.

Van Den Bosh, C. J. H. (1992). *Methods for the determination of possible damage to people and objects resulting from releases of hazardous materials (Green Book)*. Committee for the Prevention of Disasters.

Van Den Bosh, C. J. H. J. H., & Weterings, R. A. P. M. (2005). *Methods for the calculation of physical effects due to releases of hazardous materials (liquids and gases) (Yellow Book)* (3rd ed.). Committee for the Prevention of Disasters.

Veritas, D. N. (2002). *Offshore reliability data handbook* (3rd ed.) Running title: OREDA-97. Cover title: Offshore reliability data.

Woessner, J., Laurentiu, D., Giardini, D., Crowley, H., Cotton, F., Grünthal, G., Valensise, G., Arvidsson, R., Basili, R., Demircioglu, M. B., Hiemer, S., Meletti, C., Musson, R. W., Rovida, A. N., Sesetyan, K., Stucchi, M., & The SHARE Consortium. (2015). The 2013 European seismic hazard model: key components and results. *Bulletin of Earthquake Engineering, 13*(12), 3553–3596. https://doi.org/10.1007/s10518-015-9795-1

Yang, Y., Chen, G., & Reniers, G. (2020). A method for assessing propagation paths and probabilities of domino effects triggered by Natech events. *Chemical Engineering Transactions, 82*, 67–72. https://doi.org/10.3303/CET2082012

Zeng, T., Chen, G., Reniers, G., & Yang, Y. (2021). Methodology for quantitative risk analysis of domino effects triggered by flood. *Process Safety and Environmental Protection, 147*, 866–877. https://doi.org/10.1016/j.psep.2020.12.042

Chapter 4
Fire Engineering and Applications to Industrial Plant Accidents

Nicola Tondini

4.1 Introduction

4.1.1 Fire Hazards

Fires represents a significant hazard for structures. Indeed, by exposing structures to elevated temperature, the thermal impact of a fire may significantly affect their load-bearing capacity owing to material degradation. On these premises, steel structures are sensitive to the fire hazard because of high thermal conductivity and small cross-section thicknesses that entail quick temperature rise in the profiles with consequent fast loss of strength and stiffness. Since structures are for their nature loaded, consequently, fire can trigger a structural collapse. Among the most exposed to fire events, there are industrial facilities/plants that treat/stock hazardous and flammable materials and high-rise building dedicated to offices, dwellings and hotels, presenting important fire loads. These structures present higher ignition probabilities due to their function and occupancy. Ignition is usually caused by human activities, operational accidents or natural hazards such as earthquakes. It is obvious that the collapse of such structures has tremendous consequences in terms of damages to the structural environment, life and economic losses, business downtime and the release of hazardous substances can significantly affect the nearby community.

More specifically, regarding fire events previously occurred in industrial environments, Chang and Lin (2006) analysed 242 accidents implying storage tanks and identified their causes to observe that most of them were caused by lightnings, human errors and natural disasters. Zheng and Chen (2011) reviewed storage tank fire accidents occurred in China and caused mostly during human operations such as maintenance and repairs. It noted that 64% of accidents took place in petrochemical

---

N. Tondini (✉)
Department of Civil, Environmental and Mechanical Engineering, University of Trento, Trento, Italy
e-mail: nicola.tondini@unitn.it

V. Cozzani and F. Paolacci (eds.), *Management Strategies and Resilience Against NaTech Events*, CISM International Centre for Mechanical Sciences 616,
https://doi.org/10.1007/978-3-031-84485-0_4

plants. Furthermore, recent studies have also shown that industrial facilities integrating petrochemical are particularly exposed to natural hazards generating technological issues, also referred to as NaTech events (Steinberg et al. 2008; Krausmann et al. 2011a). For example, the Kocaeli earthquake in 1999 (Girgin 2011b; Barka 1999) damaged the Turpaz Ismit refinery by triggering significant fires in multiples fuel storage tanks. Storage tanks ignited one after another as a domino effect namely due to radiative heat transfer, so that all the fire in the plants took 3 d to be extinguished by the authorities. In 2011, the Fukushima nuclear power plant was seriously harmed due to the tsunami caused by the Tohoku earthquake (Anzai et al. 2011; Krausmann and Cruz 2013b). The wave namely carried off storage facilities releasing hazardous materials and causing fires. Lessons learned with the Tohoku earthquake and future measures and research needs were identified in literature (Nakashima et al. 2014; Yang 2014). Important loss of containment from storage tanks can also be provoked by Hurricanes as observed in 2005 with the Hurricane Katrina in New Orleans (Godoy 2007) and in 2017 with the Hurricane Harvey in Houston (Bernier and Padgett 2019). Probabilities of flammable materials release from industrial facilities and the subsequent ignition of the spill were studied in previous research works (Moosemiller 2011; Vílchez et al. 2011). Storage tanks and pipe-rack structures containing flammable material are critical components prone to ignition and to the loss of their containment which can subsequently ignite and trigger significant fire. That can be caused by NaTech events or by fire scenarios previously initiated.

Fires taking place in industrial plants mostly consist in very large open-air fires since facilities are located outdoors, whereas fires impacting buildings are usually initiated and grow inside the structure. The latter is defined as compartment fire, that, depending on the compartment characteristics in terms linings, geometry, openings, fire load nature and distribution, can be either ventilation-controlled or fuel-controlled. Regarding fire accidents occurred in building dedicated to offices and dwellings, numerous incidents could be observed in the last decades. Focussing on high-rise buildings, a large number of fatalities occurred in 1971, a fire occurred in the high-rise Tae Yon Hak Hotel in Seoul, Korea due to LP-gas leakage. The structure did not collapse but the fire and smokes caused 163 fatalities (Willey 1972; Craighead 2009). The Joelma building with 25 floors dedicated to offices caught fire in 1974 in São Paulo, Brazil. Fire started due to a short circuit in air conditioning system in the 12th floor and spread within the whole building in 20 min causing 179 fatalities and 300 injuries (Craighead 2009). Fire can also be triggered in buildings by earthquakes as it could be observed in California in 1989 with the Loma Prieta earthquake (Davidson 2009; Lee et al. 2008). Fires following earthquakes are usually more destructive because it can happen that fire protections and water supply are impacted by seismic actions (Covi et al. 2023). The 11th September 2001, the World Trade Center Towers collapsed due to the strike of two aircrafts which initiated fire into the entire buildings. Even though the aircraft impacts were significant, the fire was the major cause of the building collapse and caused the death of more than 2000 people (Purkiss 2007). This tragic event brought modifications in design procedures for high rise buildings by introducing into codes requirements for structural robustness and resilience (Craighead 2009). In 2005, a fire initiated in the Windsor tower in Madrid without occupants in the middle of the night and required 24 h for the firefighters to be extinguished (Fletcher et al. 2006; Menéndez and Vega 2010). The fire started in the middle of the building at the 21st floor and vertically spread up to the 32nd floor and down to the 2nd floor. More recently, in 2017, a devastating fire took place in the Grenfell Tower in London causing 71 fatalities (Mitchener 2018; McKenna et al. 2019). The main reason for which the fire could spread so rapidly, was identified as being the flammable materials composing the façade elements.

Considering devastating consequences related to fire events depicted above, the development of scientific tools quantifying fire hazard has become a high priority in the research community. The proper understanding of the impact that fires can have on structures requires the study of three governing problems characterised by different time and space scales: (i) fire development analysis; (ii) thermal analysis and (iii) mechanical analysis, as illustrated in Fig. 4.1. Figure 4.1 highlights the interactions between the governing problems. Clearly, a full-coupling approach represents a more realistic scenario, but its analysis is still highly complex (see Fig. 4.1a). For instance, the spatial scale of the thermal analysis in the structural elements is an order of magnitude smaller than the spatial scale used for the development of the compartment temperature and the mechanical response (Tondini et al. 2016). Moreover, in many practical cases, such high level of refinement is not required and, therefore, a weak-coupling approach is generally employed in the design practice. This means that the three problems are sequentially solved and the interaction is only considered in one direction, as shown in Fig. 4.1b. Indeed, the fire development influences the thermal field in the structural elements, that in turn affects the structural response.

The fire development can be modelled by means of nominal fire curves, such as the ISO 834, that are not representative of the fire phenomenon and they are used in a prescriptive approach. For instance, in petrochemical plants the use of the hydrocarbon curve may be representative of the heating in the context of the prescriptive approach. Otherwise, natural fire curves can be employed in the context of a performance-based approach, also called natural fire safety concept (NFSC) or more frequently fire safety engineering (FSE). They are obtained considering the fire characteristics, the oxygen supply and the boundary conditions. They can be derived from a few parameters, as for parametric fire curves, or from more advanced models such as zone models or even computational fluid dynamics (CFD).

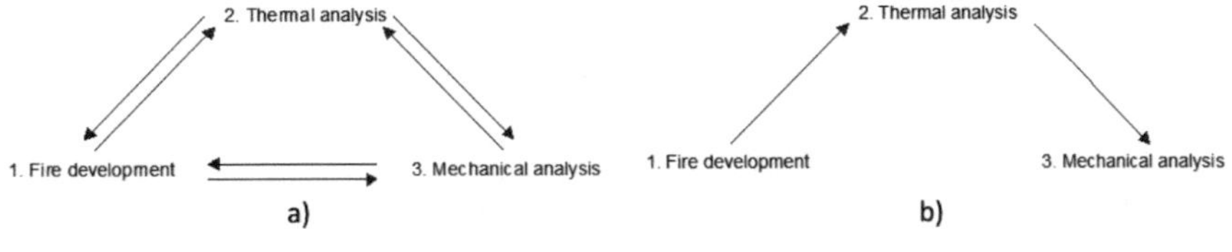

Fig. 4.1 Governing problems in structural fire engineering: **a** full-coupling approach; **b** weak-coupling approach

At the same time, the thermal analysis can be performed by means of simplified methods, such as the lumped mass concept suitable for steel structures, or by numerically solving the transient heat conduction equation according to the finite element (FE) method.

Given the evolution of the temperature in the structural elements, the mechanical response can be computed by means of tabulated data, as for member analysis of concrete structures subjected to the ISO 834 nominal fire curve (EN1991-1-2 2004), or by simple calculations in accordance with the standards (EN1993-1-2 2005) or by advanced calculation models developed, for instance, according to the finite element method, when part of or the whole are modelled.

In the following, the description of fire models and thermo-mechanical analysis specifically applied to steel structures that are commonly used in industrial plants is provided.

4.1.2 Fire Models

Based on the fire events depicted in Sect. 4.1.1, the scientific community has undertaken the development of models to study the effect of fire on structures. For that purpose, numerous experimental campaigns were conducted. Analytical and numerical fire models were then developed and calibrated against experimental results. These models aim to quantify the effects in terms of thermal loadings, i.e. temperatures and/or heat fluxes, that represent the boundary conditions to solve the transient heat conduction equation. Typically, the most representative boundary conditions of the effect of a fire are due to convection and radiation, as illustrated in Fig. 4.2. In some cases, one heat transfer mode is predominant. For instance, in external fires occurring in industrial plants, when the structure is not engulfed into the fire, radiation is the governing heat transfer mechanism.

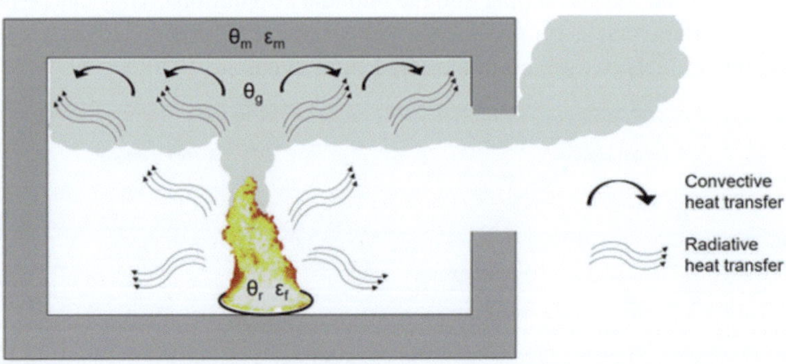

Fig. 4.2 Heat transfers from a fire within a compartment

The net heat flux received at the surface of a structural element exposed to fire is defined in Eq. (4.1) as the sum of the net convective and radiative heat fluxes. In fire engineering, the heat flux transferred to the surface by convection is expressed through Eq. (4.2). The net radiative heat flux results from the difference between the absorbed portion of the incident heat flux and the heat flux re-emitted by the surface, as expressed in Eq. (4.3). The effective radiation temperature of the fire environment taken into account for the incident radiative heat flux calculation is defined in Eq (4.4). Eventually, Eq. (4.5) defines explicitly the net heat flux received by the surface of a structural member exposed to fire.

$$\dot{q}_{net} = \dot{q}_{con} + \dot{q}_{rad} \tag{4.1}$$

with

$$\dot{q}_{con} = \alpha_c \left(\theta_g - \theta_m\right) \tag{4.2}$$

$$\dot{q}_{rad} = \varepsilon_m \, \sigma \, \theta_r{}^4 - \varepsilon_m \, \sigma \, \left(\theta_m + 273\right)^4 \tag{4.3}$$

$$\theta_r{}^4 = \sum_i \phi_i \, \varepsilon_{f,i} \left(\theta_{f,i} + 273\right)^4 \tag{4.4}$$

$$\dot{q}_{net} = \alpha_c \left(\theta_g - \theta_m\right) + \varepsilon_m \, \sigma(\theta_r{}^4 - (\theta_m + 273)^4) \tag{4.5}$$

where

$\dot{q}_{net}$	net heat flux received by the surface exposed to fire	(W/m^2)
$\dot{q}_{con}$	net heat flux transferred to the surface through convection	(W/m^2)
$\dot{q}_{rad}$	net heat flux transferred to the surface through radiation	(W/m^2)
α_c	coefficient of convection	(W/(°C.m^2))
θ_g	temperature of the gas surrounding structure	(°C)
θ_m	structural member temperature	(°C)
θ_r	effective radiation temperature of the fire environment	(K)
ε_m	emissivity of structural member	(–)
σ	Stephan Boltzmann constant	(W/m^{2}K^4)
ϕ_i	configuration factor between structural member and a fire i	(–)
$\varepsilon_{f,i}$	emissivity of a fire i	(–)
$\theta_{f,i}$	temperature of a fire i	(°C)

In EN1991-1-2 (EN1991-1-2 2004), the net heat flux received by a structural element when exposed to a fire is defined by Eq. (4.6), where the first term is the convective part and the second term is the radiative part of the heat flux, which presents a discrepancy with respect to Eq. (4.5). Indeed, it can be noted that Eq. (4.6) is equivalent to Eq. (4.5) when a fire is assumed large and fully developed so that ϕ and ε_f are equal to 1 and θ_r is equal to the gas temperature.

$$\dot{q}_{net} = \alpha_c \left(\theta_g - \theta_m\right) + \phi \, \varepsilon_f \, \varepsilon_m \, \sigma \left[(\theta_r + 273)^4 - (\theta_m + 273)^4\right] \tag{4.6}$$

where

ϕ	configuration factor between structural member and a fire	(–)
ε_f	emissivity of fire taken as 1	(–)
θ_r	radiation temperature of the fire environment	(°C)

In the next subsections, the most common fire models used in the design practice will be described. These models are also provided in EN1991-1-2 (2004).

4.1.2.1 Nominal Fire Curves

Nominal fires refer to nominal temperature-time curves defined in the context of a prescriptive approach. Thus, they are not representative of the real fire phenomenon. Two nominal temperature-time curves are frequently used in fire engineering: the standard curve and the hydrocarbon curve. The standard heating or fire curve, also referred to as ISO 834, represents the gas temperature evolution with time within a whole compartment where a fully-developed fire develops. That curve was defined to address fires that could occur in buildings with fire load mainly characterised by cellulosic material. The standard curve is also taken as the reference curve when evaluations and comparisons of products are made in terms of fire resistance. Indeed, that curve has been adopted to assess the fire resistance of structural elements or the insulating efficiency of fire protections. That standard curve is defined in Eq. (4.7) where θ_g and t refers to the gas temperature in °C and time in minutes, respectively.

$$\theta_g(t) = 20 + 345 \log_{10} (8t + 1) \tag{4.7}$$

The hydrocarbon curve is used to represent the thermal effects of fire resulting from fuel made of hydrocarbons, e.g. gasoline, kerosene etc. That curve was defined to take into account the severity of such fires. Critical components within industrial plants can be designed against fire by considering this curve. The hydrocarbon curve is defined in Eq. (4.8) where θ_g and t also refers to the gas temperature in °C and time in minutes, respectively.

$$\theta_g = 20 + 1080(1 - 0.325e^{-0.167t} - 0.675e^{-2.5t}) \tag{4.8}$$

Nominal temperature-time curves are depicted in Fig. 4.3. It can be observed that the hydrocarbon curve entails a more severe temperature increase than the standard curve, reaching the plateau of 1100 °C in 30 min, whilst the standard curve reaches this value after about 180 min. However, the latter keeps increasing whereas the former reaches a maximum temperature of 1100 °C.

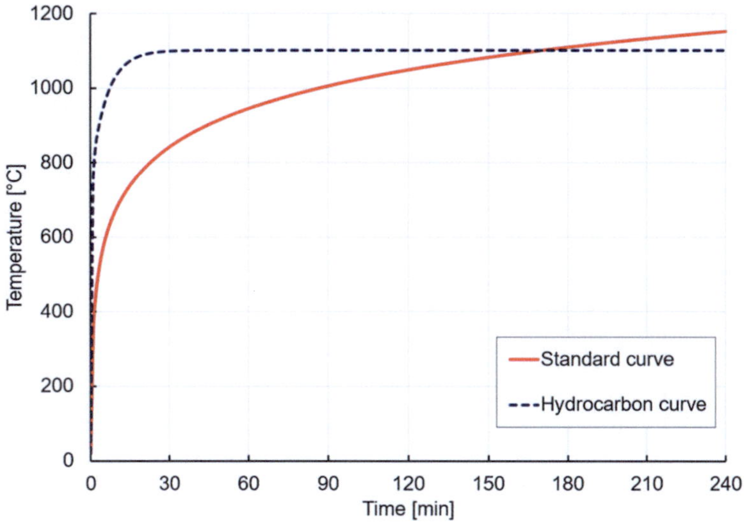

Fig. 4.3 Nominal temperature-time curves

4.1.3 Parametric Fire Curves

Parametric temperature-time curves represent the gas temperature evolution with time within a compartment as a function of the compartment properties. They represent a fully-developed fire. The methodology to develop such curves is further detailed in Annex A of EN1991-1-2 (2004). These curves allow for the cooling phase and they are defined knowing the geometry, the compartmentation materials and the fuel load density of the compartment. These three parameters govern the development of a fire within a compartment and therefore influence the gas temperature evolution with time. As depicted in Fig. 4.4, compartment fires are characterised by heating and cooling phases. In the heating phase the temperature increases with time in a nonlinear way until it reaches a maximum temperature θ_{max} at time t_{max}. In the cooling phase the temperature linearly decreases with time. More precisely, the heating phase of parametric temperature-time curves is defined in Eq. (4.9) where θ_g is the gas temperature in °C and t^* is the expanded time, defined as the time t in hour multiplied by the expansion coefficient Γ expressed in Eq. (4.11). That coefficient is function of the opening factor O and the parameter b. Opening factor O quantifies the ventilation of the compartment and is defined in Eq. (4.12) where A_v is the opening area, A_t is the total area of the boundaries (walls, floor and ceiling, including openings) and h_{eq} is the weighted average of the opening heights. The b parameter quantifies the capacity of the compartment boundaries to absorb heat and is defined in Eq. (4.13) where c_p is the specific heat in J/kgK at constant pressure, ρ is the density in kg/m^3 and λ is the thermal conductivity in W/mK of the material constituting the enclosure. When compartment boundaries are defined with different

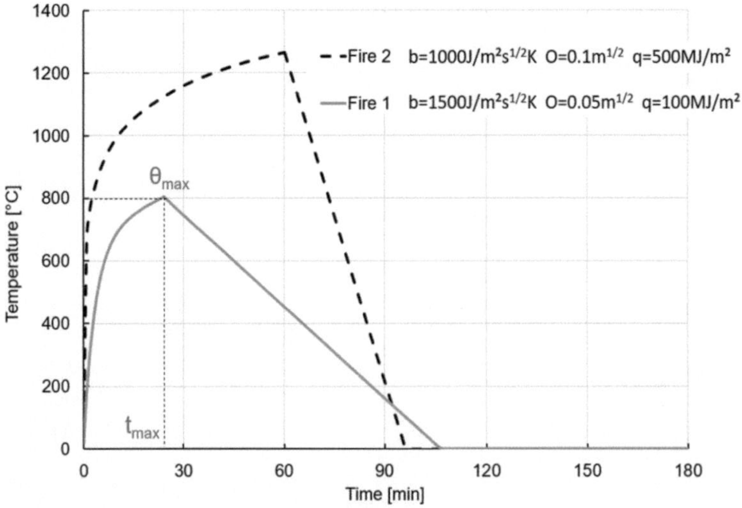

Fig. 4.4 Parametric temperature-time curves

materials, parameters b_i can be defined for each material and a mean parameter b weighted by the material areas ai must be considered. The duration of the heating phase t_{max} is defined in Eq. (4.14) as a function of the opening factor O and the design fire load density $q_{t,d}$ in MJ/m^2. That design value is related to the total area of the boundaries A_t, as defined in Eq. (4.15), considering the fire load density $q_{f,d}$, depending on the compartment occupancy and related to the floor area A_f. Three compartment fires presenting different parameters b, O and $q_{t,d}$, are compared in Fig. 4.4.

$$\theta_g = 20 + 1325(1 - 0.324e^{-0.2t^*} - 0.204e^{-1.7t^*} - 0.472e^{-19t^*}) \tag{4.9}$$

$$t^* = \Gamma\, t \tag{4.10}$$

$$\Gamma = \left(\frac{O/0.04}{b/1160}\right)^2 \tag{4.11}$$

$$O = A_v\sqrt{h_{eq}}/A_t \tag{4.12}$$

$$b = \sqrt{c\rho\lambda} \tag{4.13}$$

$$t_{max} = 0.0002\, q_{t,d}/O \tag{4.14}$$

$$q_{t,d} = q_{f,d}\, A_f/A_t \tag{4.15}$$

4.1.4 Zone Models

Zone models, such as Ozone (Cadorin et al. 2003) and CFAST (Peacock et al. 2015) among the others, are advanced models defining the gas temperature evolution with time within a compartment by solving partial differential equations governing the exchange of mass and heat between the compartment and the surrounding environment. These models are named "zone" models because they can represent the gas temperature in a compartment by dividing it into one or several zones. When a fire within a compartment is in the ignition phase, i.e. localised and confined, it is worth considering different zones within compartment, i.e. a hot zone at upper level of the compartment and a cold zone at the lower level of the compartment. For fully-developed fires in a compartment, it is more relevant to consider one zone assuming that the gas temperature is uniform within the compartment. Zone models yield temperature-time curves based on three characteristics defining a compartment, i.e. ventilation, boundaries/compartmentation and fuel. Parametric fire curves quantify these characteristics with lumped parameters O, b and $q_{t,d}$, whereas zone models consider them in a more accurate manner. Ventilation is function of the openings taken as input data with their actual dimensions and position within the compartment. Opening variation in time can also be introduced in zone models which can simulate the effects of smoke extractors, automatic openings or glass breaking due to high temperatures. Accurate compositions and thicknesses are considered for the layers defining walls, floor and ceiling. Therefore, heat transfers in all boundaries can be quantified through conduction and radiation for cavities. Regarding the representation of the fuel in the compartment, two aspects are taken into account; the type and amount of fuel and the rate of heat release (RHR) corresponding to that fuel. Based on these data, zone models can assess whether the fire is "fuel controlled" or "ventilation controlled" and can adapt the combustion model accordingly which, influences the gas temperature evolution with time. Due to their complexity, zone models require the use of software solving numerically the mass and heat differential equations for given compartment configurations, as for example the software OZone (Cadorin et al. 2003) developed in the University of Liège or the software CFAST (Peacock et al. 2015) developed by the NIST.

4.1.5 Field Models

Field models, also referred to as computational fire dynamics CFD models, offer a high modelling level of the fire behaviour by solving Navier-Stokes partial differential equations of the fluid flow. Considering a fire and its environment in three dimensions, field models discretise that environment with a very large number of control cells in which the partial differential conservation equations are solved iteratively considering a time step previously defined. The application of field models requires the knowledge of physical and chemical reactions relative to the fire combustion processes, as well as the boundary conditions of the fire environment which can impact the fire behaviour. These models accurately address the modelling of any fire scenario and represent environments with complex geometries. In particular, they allow an efficient prediction of air and smoke movements, facilitating the identification of ventilation issues within buildings. The application of field models is very demanding in terms of calculation time and requires therefore high computational capacity and appropriate software, such as Fire Dynamic Simulator (FDS) (McGrattan et al. 2013). Field models can be used to model localised fires or jet fires occurring in industrial fire accidents.

4.1.6 Localised Fires

Localised fire models were developed to address scenarios where a fire takes place on a limited compartment area. For compartment fires, such scenarios typically correspond to the fire ignition phase before the flash over, if it occurs. Fire accidents in industrial plants owing to leakage from a tank or a pipe can ignite pool fires that can be modelled by means of localised fire models. For structural members engulfed in the fire area, the heat transfer between the flame and the member is governed by convection and radiation, whereas for members located outside the fire, the main heat transfer mechanism is radiation. Thus, the definition of localised fire models able to predict thermal radiation is paramount. In this respect, several models were developed to study the thermal radiation from hydrocarbon pool fires based on experimental observations (Rew et al. 1997; Beyler 2016; Shokri and Beyler 1989; Heskestad 1984; Babrauskas 1983; Thomas 1963; Mudan et al. 1984; Ufuah and Bailey 2011; Ding et al. 2014). In more detail, Rew and Hulbert (Rew et al. 1997) defined a mathematical semi-empirical model to quantify the thermal radiation from hydrocarbon pool fires. Shokri and Beyler (1989, 2016) developed a method to predict the radiative heat flux emitted from large pool fires and received by a target outside the flame, by exploiting the Heskestad correlations (Heskestad 1984) to characterise the flame length. This model gathers the research works achieved by Babrauskas (1983), Thomas (1963), and Mudan (1984) for the definition of the mass burning rate, the flame length, and the surface emissive power from the flame, respectively. The Shokri and Beyler model, the Mudan model and the point source models are compared, and significant differences are discussed by Ufuah and Bailey (2011), Li-bin et al. (2014) and Beyler (2016). More recently, in the framework of a European research project, the LOCAFI analytical model was developed to compute the radiative heat flux received by vertical structural elements from a pool fire (Francis et al. 2018; Tondini et al. 2019; Hanus et al. 2016; Tondini and Franssen 2017). In Annex C of EN 1991-1-2 (EN1991-1-2 2004) the Hasemi model is provided to compute the heat flux at the ceiling level when the flame impacts it. Moreover, localised fire models are also appropriate to represent outdoor fires and fires taking place in very large compartment where flash over is unlikely to occur. Fire models, defining a temperature-time curve of the gas surrounding a structure, are convenient to deal with since that temperature can be inserted in Eq. (4.6) and the net heat flux impinging the structure is straightforwardly determined. However localised fire models aim to characterise the fire severity by defining directly the heat fluxes emitted from the fire source and impinging surrounding structures. For structural members engulfed in the fire area, the heat transfer between the flame and the member is governed by convection and radiation. Whereas for members located outside the fire, the main heat transfer mechanism is radiation. Advanced fire models, such as field models described in Sect. 4.1.2, can be used to analyse localised fires by computing heat fluxes emitted from fire. Nevertheless, considering their complexity level, several analytical models were developed in the last decades to characterised localised fires behaviour in a simplified manner. More precisely, these analytical models aimed to quantify the flame geometry and the radiative heat fluxes emitting from fire, based on a dimensionless modelling and experimental data. Most analytical models represent localised fires with a solid flame model by adopting a specific shape for the flame geometry, such as a circular cylinder, an elliptical cylinder or a cone (Rew et al. 1997; Ufuah and Bailey 2011; Tondini et al. 2019). Considering a structural element located beside a fire, analytical models determine the radiative heat flux impinging this element as the sum of the radiative heat fluxes emitted from all the surfaces defining the solid flame. For each surface a configuration factor ϕ has to be computed, which can be achieved analytically or numerically, depending on the shape adopted for the flame. Even though solid flame models are burdensome in terms of calculation, they provide accurate representation of the heat flux associated with a larger field of application. Among existing localised fire models, the LOCAFI model, recently develop within the European LOCAFI Project (Francis et al. 2018), appeared particularly appropriate for simulating the thermal impact from a pool fire occurring in an industrial plant. The LOCAFI model was developed and calibrated by Tondini et al. (2019) based on experimental tests (Tondini and Franssen 2017) and numerical analyses performed with CFD models. The LOCAFI model adopts a solid flame concept to characterise the localised fire and the parameters relative to the model are illustrated in Fig. 4.5. An analytical and a numerical model were developed. The former is more suitable for hand calculation, whereas the latter to be implemented in a software, as shown in Fig. 4.6. Fuel material liquid or solid distributed on a reduced area and source of a localised fire, has to be represented with an equivalent circular area. Annex C of EN1991-1-2 (2004) provides the correlations that allows to characterise the LOCAFI model in terms of flame height and temperature. Hence, the length of the flame *Lf* is expressed in Eq. (4.16), as defined by Heskestad (1984), as a function of the fire diameter and the RHR. The temperature along the vertical axis of the flame can be expressed with Eq. (4.17) as long as the localised fire takes place in open air or does not impact a compartment ceiling. That temperature is function of the height z, the convective part of the RHR and the virtual origin of the fire source, both defined in Eqs. (4.18) and (4.19).

$$L_f = 0.0148\ Q^{0.4} - 1.02\ D \qquad \text{(m)} \tag{4.16}$$

Fig. 4.5 LOCAFI model: **a** Localised fire; **b** Conical flame representation

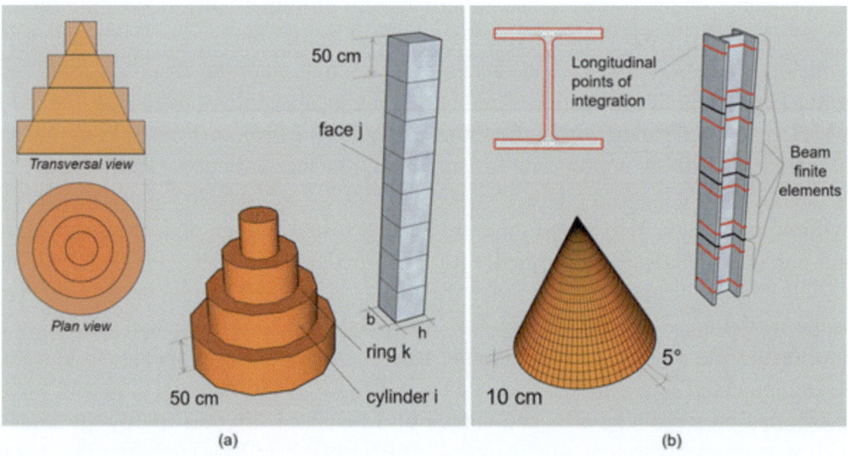

Fig. 4.6 LOCAFI models: **a** analytical model; **b** Numerical Model

where

L_f	length of the conical flame	(m)
Q	rate of heat release (RHR)	(W)
D	fire diameter	(m)

$$T(z) = 20 + 0.25\ Q_c^{2/3}\,(z - z_0)^{-5/3} \le 900 \qquad (°C) \qquad (4.17)$$

with

$$Q_c = 0.8Q \qquad \text{(W)} \tag{4.18}$$

$$z_0 = 0.00524Q^{0.4} - 1.02D \qquad \text{(m)} \tag{4.19}$$

where

$T(z)$	temperature along the flame axis	(m)
Q_c	convective part of the RHR	(W)
z	height considered along the flame axis	(m)
z_0	virtual origin of the fire source	(m)

Based on these correlations, a conical shape and the temperature evolution along the flame axis, can be defined considering a fire diameter and a type of fuel for which the corresponding RHR can be determined, e.g. with Annex E of EN1991-1-2 (2004). Once a localised fire is defined, incident radiative heat fluxes can be computed for any external element considered. Equation (4.20) defines the incident radiative heat flux emitted by a surface A and impinging a surface B.

$$\dot{q}_{A\to B} = \phi_{A\to B}\, \varepsilon_A\, \sigma\, (\theta_A + 273)^4 \qquad (\text{W/m}^2) \tag{4.20}$$

where

$\dot{q}_{A\to B}$	is the radiative heat flux	(W/m^2)
$\phi_{A\to B}$	is the configuration factor between surfaces A and B	(–)
ε_A	is the surface A emissivity	[-]
σ	is the Stephan Boltzmann constant	$(\text{W/m}^2\text{K}^4)$
θ_A	is the surface A temperature	(K)

That radiative heat flux is represented in Fig. 4.5b where the surface A is an element discretizing the conical fire and the surface B is an element discretizing a structural steel member. The configuration factor varies between 0 and 1 as a function of the sizes and orientations of surfaces A and B and the distance separating them. The calculations of the numerous configuration factors between elements discretising localised fire and structural members can be highly complex depending on the adopted discretisation level. It is clear that, in general, localised fires determine a certain degree of thermal gradient within the cross section and along the member, that should be taken into account in the subsequent thermo-mechanical analysis (Kamikawa et al. 2003; Ferraz et al. 2016; Possidente et al. 2024).

4.1.7 Thermo-Mechanical Analysis of Steel Structures

This section introduces the analysis methods adopted to quantify the thermo-mechanical effects that fire has on steel structures. Heat transfer between fire and structural member is mainly governed by convection and radiation whereas heat transfer within the cross-sections of steel members is governed by conduction. In general, in order to obtain the evolution of the thermal field with time within a steel cross-section, analytical or numerical methods are adopted. It is worth pointing out that, when steel profiles such as hot-rolled profiles, are used, a 2D thermal analysis of the cross section generally suffices, which means that conduction along the length of the member is neglected. Consequently, in the finite element context, thermo-mechanical beam finite elements, e.g. (2020), are employed, which for steel pipe-racks can be a good approximation, whereas for steel tanks, thermomechanical shell elements, e.g. Possidente et al. (Possidente et al. 2019), are preferable (Pantousa 2018). EN1993-1-2 (2005) provides simplified analytical models, both for unprotected and protected steel elements based on a mass lumped approach, considering a uniform temperature within a steel member cross section. This is a good approximation because structural steel profiles are characterised by components with low Biot numbers (typically <0.1), which entail low temperature gradients within the thickness. It has been shown that for heavily protected steel members the formula given by the EN 1993-1-2 can provide inaccurate results and, in this respect, Possidente et al. (2023b) and Possidente and Tondini (2023) developed and validated an improved formulation. Finite element models can be employed to accurately study the temperature field within cross-sections by solving the heat conduction equation by considering appropriate boundary conditions. The behaviour study of a steel structure exposed to fire is usually divided into two parts. The first part consists in determining the steel temperature within the whole structure depending on thermal actions. The second part consists in predicting the structural behaviour in terms of displacements and deformations considering the mechanical loadings and the evolution of the steel mechanical properties with temperature. For steel structures, it is possible to employ simple calculation models for member analysis in the time domain, strength domain or temperature domain, as for EN 1993-1-2 (2005). In numerical simulations, usually part of the structure or the whole structure is modelled and thermal and mechanical analyses can be either performed in a coupled or de-coupled manner. Coupled thermo-mechanical analyses consider simultaneously both thermal and mechanical loadings with their mutual effects. Conversely, de-coupled thermo-mechanical analyses consider first thermal analyses and then the mechanical ones based on the thermal results. In this context, it is important to be able to capture nonlinear behaviour, both materially and geometrically, given the material property degradation and large displacements associated with the behaviour of steel structural systems. When analysing the behaviour of steel structures in fire, suitable steel material properties have to be considered. In particular, EN 1993-1-2 (2005) provides guidance both for thermal and mechanical properties at elevated temperature and they are summarised herein. As mechanical properties of steel are reduced at elevated temperature, the strain-stress relationship is modified as illustrated in Fig. 4.7. The effective yield strength at 2% total strain is considered. Key parameters reduced at elevated temperature are yield strength f_y, proportional limit f_p and Young modulus E. They must be multiplied by reduction factors $k_{y,\theta}$, $k_{p,\theta}$ and $k_{E,\theta}$, respectively, as expressed in Eqs. (4.21)–(4.23). These

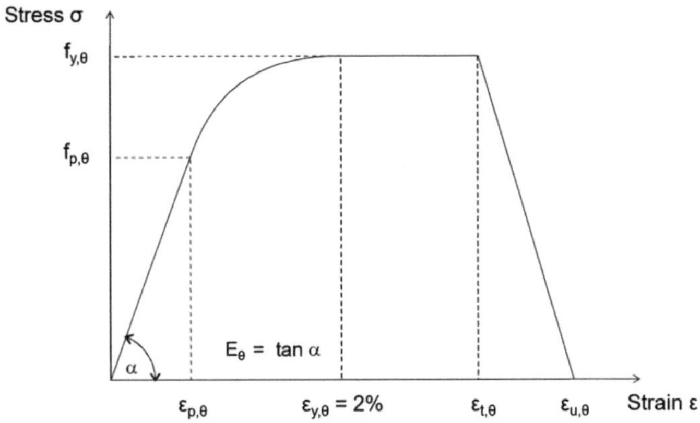

Fig. 4.7 Stress-strain relationship for carbon steel at elevated temperatures (EN1993-1-2 2005)

reduction factors represent the degradation of steel with temperature as depicted in Fig. 4.8.

$$f_{y,\theta} = k_{y,\theta}\, f_y \tag{4.21}$$

$$f_{p,\theta} = k_{p,\theta}\, f_y \tag{4.22}$$

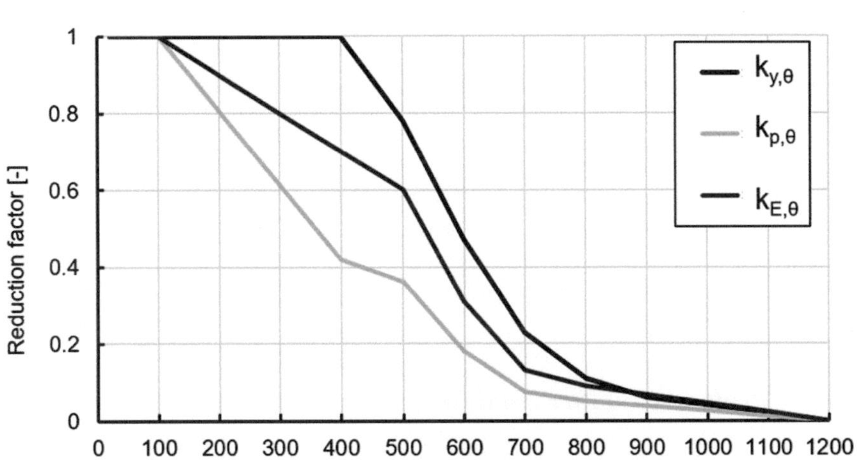

Fig. 4.8 Reduction factors for yield strength limit, proportional limit and Young modulus at elevated temperatures

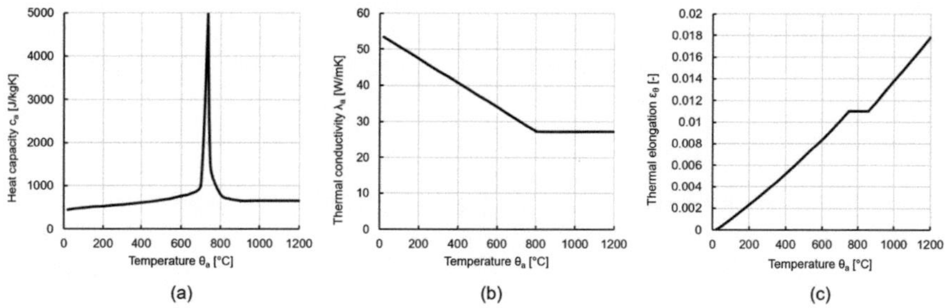

Fig. 4.9 Thermal properties of steel at elevated temperatures: **a** Heat capacity c_a; **b** Thermal conductivity λ_a; **c** Thermal elongation ε_θ

$$E_\theta = k_{E,\theta}\, E \tag{4.23}$$

Thermal properties of steel varying with temperature are heat capacity c_a, thermal conductivity λ_a and thermal elongation ε_θ. Heat capacity variation with temperature is defined in Eq. (4.24) and depicted in Fig. 4.9a. Thermal conductivity variation with temperature is defined in Eq. (4.25) and depicted in Fig. 4.9b. Thermal elongation variation with temperature is defined in Eq. (4.26) and depicted in Fig. 4.9c, where θ_a is the steel temperature, L is the length of any steel member at 20°C and ΔL its corresponding thermal dilatation.

$$C_a = \begin{cases} 425 + 7.73\ 10^{-1}\ \theta_a - 1.69\ 10^{-3}\ \theta_a^2 + 2.22\ 10^{-6}\ \theta_a^3 \\ 666 + 13002\ /\ (738 - \theta_a) \\ 545 + 17820/\ (\theta_a - 731) \\ 650 \end{cases} \text{for} \begin{array}{l} 20 \le \theta_a \le 600\ °\text{C} \\ 600 < \theta_a \le 735\ °\text{C} \\ 735 < \theta_a \le 900\ °\text{C} \\ 900 < \theta_a \le 1200\ °\text{C} \end{array} \ (\text{J/kgK}) \tag{4.24}$$

$$\lambda_a = \begin{cases} 54 - 3.33\ 10^{-2}\theta_a \\ 27.3 \end{cases} \text{for} \begin{array}{l} 20 \le \theta_a \le 600\ °\text{C} \\ 800 < \theta_a \le 1200\ °\text{C} \end{array} \ (\text{W/mK}) \tag{4.25}$$

$$\varepsilon_\theta = \frac{\Delta L}{L} = \begin{cases} 1.2\ 10^{-5}\ \theta_a + 0.4\ 10^{-8}\ \theta_a^2 - 2.416\ 10^{-4} \\ 1.1\ 10^{-2} \\ 2\ 10^{-5}\ \theta_a - 6.2\ 10^{-3} \end{cases} \text{for} \begin{array}{l} 20 \le \theta_a \le 750\ °\text{C} \\ 750 < \theta_a \le 860\ °\text{C} \\ 860 < \theta_a \le 1200\ °\text{C} \end{array} \ [-] \tag{4.26}$$

4.1.8 Probabilistic Fire Analysis

In this section, some common design approaches are reviewed. In particular, probabilistic fire analysis and derivation of fragility functions are described.

Designing a structure consists in evaluating structural capacity and structural demand in terms of loads. In fire engineering, these loads usually refers to mechanical and thermal actions impacting structures. Both structural capacity and demand present random nature. Structural capacity refers to the ability of a structure to carry loads which relies on materials, connections and boundary conditions. Structural demand refers to load combinations and external actions applied on the structure usually depending on structural environment and assumptions made by stakeholders. It worth noting that randomness is often greater for structural demand than capacity. Structural design aims to ensure that demand D does not exceed bearing capacity C, as expressed in Eq. (4.27). Two approaches can be adopted to quantify structural demand and structural capacity: semi-probabilistic (Eurocode-type approach) and probabilistic approaches.

$$D < C \tag{4.27}$$

Semi-probabilistic approach consists in deterministically comparing the structural capacity C and structural demand D described by characteristic values obtained from probabilistic analysis and translated into design values with the use of partial safety factors (EN1990 2005).

A probabilistic approach consists in characterising structural capacity C and structural demand D with probability functions to properly take randomness into account. Fragility functions are convenient tools when decisions are to be taken in structural design considering uncertainties. More specifically, for a given structure exposed to fire, a fragility function computes the probability that a structure reaches a predefined damage state (e.g. beam deflection level, interstorey drift ratio, critical temperature, etc.) depending on an intensity measure characterising the fire (e.g. fire load, fire distribution, etc.).

A fragility function expresses the probability that an engineering demand parameter (EDP) exceeds a structural limit state (LS) as a function of an intensity measure (IM). That probability is usually written in the following form

$$P(EDP > LS|IM) \tag{4.28}$$

The aim is to develop fire fragility functions to be used in a probabilistic framework that will serve as a means for practitioners to probabilistically assess/design a structure, e.g. a steel pipe-rack in an industrial plant, subjected to a particular hazard, e.g. localised fires. To build up fragility functions and to define the probability expressed in Eq. (4.28), it is necessary to develop a probabilistic demand model. In literature, several probabilistic seismic demand models (PSDM) can be found. Hence, appropriate PSDM can be developed by means of cloud analysis (CA), incremental dynamic analysis (IDA) and multiple stripe analysis (MSA). Shome and Cornell (1998) employed CA by assuming that EDPs follow a lognormal distribution when conditioned on IM. Parameters characterising this distribution can then be used to define a fragility function. Cornell and Vamvastikos (2002) derived fragility functions by using non-linear dynamic analyses to scale IMs until the resulting EDP exceeds LS. IDA is particularly convenient in the earthquake field where scaling seismic records is adequate, whereas for fire applications scaling the fire properties can lead to unrealistic fire scenarios. Gardoni et al. (2002) developed a methodology to construct probabilistic capacity models of structural components. Baker (2015) developed MSA that aims to represent for each IM considered, a distinct stripes of the resulting EDPs value from which, considering a LS, a probability of exceedance can be computed. Each IM is then assigned with a probability value and a fragility curve can be defined to fit these values. These methodologies are discussed and compared by Mackie and Stojadinovic (2005). Several examples of PSDM development and application can be found in literature. Hence, Phan et al. (2017) investigated the seismic behaviour of elevated steel storage tanks supported by reinforced concrete columns, by developing PSDMs and deriving fragility curves using both IDA and CA. Mackie and Stojadinovic (2003) developed a PSDM for highway overpass bridges and highlighted the importance of using efficient EDP-IM pairs. Tondini and Stojadinovic (2012) developed a PSDM and derive fragility functions for curved reinforced concrete bridges adopting CA. Bernier and Padgett (2019) studied the behaviour of aboveground storage tanks when exposed to multi-hazard, such as surge, wave and wind loads, by developing fragility functions. Regarding fire engineering field, only few probabilistic fire demand models PFDM have been developed nowadays. Lange et al. (2014) employed the performance-based earthquake engineering (PBEE) framework developed by the Pacific earthquake engineering research centre (PEER), and adapted it for fire engineering applications. Gernay et al. (2016, 2019) developed a methodology to derive fragility functions addressing steel buildings exposed to compartment fires. By quantifying structural demand and capacity with probability distribution in the temperature domain, fragility curves were derived by convolution of the distributions. Shrivastava et al. (2019) also adopted a PBEE framework to study compartment fires. By adapting MSA as PFDM, different intensity measures characterising temperature time-curves were compared based on their efficiency to represent the fire severity.

4.1.9 Probabilistic Fire Analysis Applied to Industrial Plants

In industrial and petrochemical plants, piping systems are used to transport flammable material, liquid or gas fuel, on long distances. They are usually supported by pipe-rack steel structures, passing between other industrial components, e.g. tanks, distillation towers, etc. These structures usually unprotected due to their significant length, are components among the most exposed to natural hazards, such as earthquakes, hurricane or tsunamis. It appears in plants that under certain conditions, these events can trigger severe structural damages, leading a pipe or a tank to lose its containment. Furthermore, petrochemical plants are locations where a releasing flammable material is more likely to ignite than anywhere else due to activities and operations taking place in industrial environments. Even if the probability of occurrence is low, the consideration of such scenarios cannot be ignored, given their catastrophic consequences. Numerous studies have recently highlighted the exposure of petrochemical plants piping systems to natural hazards generating technological issues, also named NaTech events.

Uehara (1991) focused on fire safety assessments in petrochemical plants by prescribing a method to evaluate the potential dangers of fire accident within facilities storing flammable material. Pantousa (2018) investigated the behaviour of steel storage tanks when exposed to the thermal radiation from surrounding burning tanks. Shu and Chong [66] simulated fires and explosions within a Naphtha-cracking plant and evaluated the resulting risks through a quantitative risk analysis. They namely considered an existing steel pipe rack as case study and delivered performance-based seismic analyses. Campedel et al. (2008b) developed a procedure to assess industrial risk due to NaTech events, with the use of a quantitative risk analysis.

Specifically, and as an example, the application of a probabilistic fire analysis to investigate the thermo-mechanical response of a steel pipe-rack entails the selection of a meaningful number of fire scenarios of pool fire located beside it with the aim to build a probabilistic fire demand model from which deriving fire fragility functions. For that purpose, a pool fire model must be selected, allowing the simulation of a burning tank and a localised fire resulting from leakage and loss of containment. It is clear that the investigation of the causes of ignition of a pool fire in an industrial or petrochemical plant represents a key factor to highlight that the selected scenarios, though with low probability, are plausible. In this respect, Gottuk and White (2016) provided methods to evaluate the occurrence of liquid fuel leakage and the consequent growth and the size of the potential resulting fire. The 'Yellow Book', developed by the committee for the prevention of disasters by hazardous materials (2005), provides a method to quantify the mass flow rate resulting from a fuel leakage through a hole in a tank wall or in a pipe considering the pressure loss.

The use of fragility functions, considering efficient intensity measures to characterise the fire severity, allows the mitigation of fire risks when petrochemical plants are designed. The efficiency and sufficiency of IM was investigated in earthquake engineering by Luco and Cornell (2007) and in fire engineering by Shrivastava et al. (2019). Ebrahimian et al. (2015) proposed an alternative manner to evaluate the sufficiency of IM by computing the relative sufficiency between two potential IMs. This measure aims at evaluating the sufficiency of an IM comparatively with another one by considering the amount of information gained (on average) about the structural response. The relative sufficiency measure between IM_1 and IM_2 is defined in Eq. (4.29) and is expressed in unit of bits of information.

$$I\left(EDP\left|IM_2\right|IM_1\right) \approx \frac{1}{n}\sum_{i=1}^{n}\log_2\frac{p[EDP = EDP_i|IM_2]}{p[EDP = EDP_i|IM_1]} \tag{4.29}$$

EDP_i are the EDP values calculated for each of the n fire scenarios and the probability distribution function (PDF) $p[EDP|IM]$ is defined in Eq. (4.30), considering the dispersion, and where $\phi\,(*)$ is the standard Gaussian PDF.

$$p\left[EDP = EDP_i | IM\right] = \frac{1}{EDP_i \cdot \beta_{EDP|IM}} \phi\left(\frac{\ln\left(EDP_i / (a\ IM_i^b)\right)}{\beta_{EDP|IM}}\right) \quad (4.30)$$

IM_2 is more sufficient than IM_1 if IM_2 provides more information about the structural response (EDP) than IM_1, which is the case when $I(EDP|IM_2|IM_1)$ is positive. Symmetrically, IM_2 is less sufficient than IM_1 when $I(EDP|IM_2|IM_1)$ is negative.

Application of the probabilistic approach to a steel pipe-rack belonging to a petrochemical plant subjected to localised fires was described in (2021, 2023a). The reference steel pipe-rack was taken from an existing petrochemical plant located in Italy, as depicted in Fig. 4.10, which was modelled with beam finite elements, as illustrated in Fig. 4.11. Several fire scenarios were considered to build up a probabilistic fire demand model that takes into account uncertainties in the fire (size, fuel type, distance from the structure) (Randaxhe et al. 2021). In Possidente et al. (2023a) the uncertainties related to the steel mechanical properties and whether the fires were caused by either leakage from a tank or from a pipe were included. The LOCAFI model was applied as localised fire model. Different IM and EDP were investigated and both the CA and MSA were applied to develop the PFDM.

In particular, in the CA, all fire scenarios are run and, based on thermo-mechanical results, a cloud of points is plotted on an EDP vs IM chart. Assuming that the EDP follows a lognormal distribution when conditioned on the IM, which is a common assumption, it is possible to write the median EDP, i.e. $E\hat{D}P$, as expressed in Eq. (4.31). Therefore, the conditional median of EDP given IM is linear in log-log space, as shown in Eq. (4.32), whereas the conditional dispersion of EDP given IM is

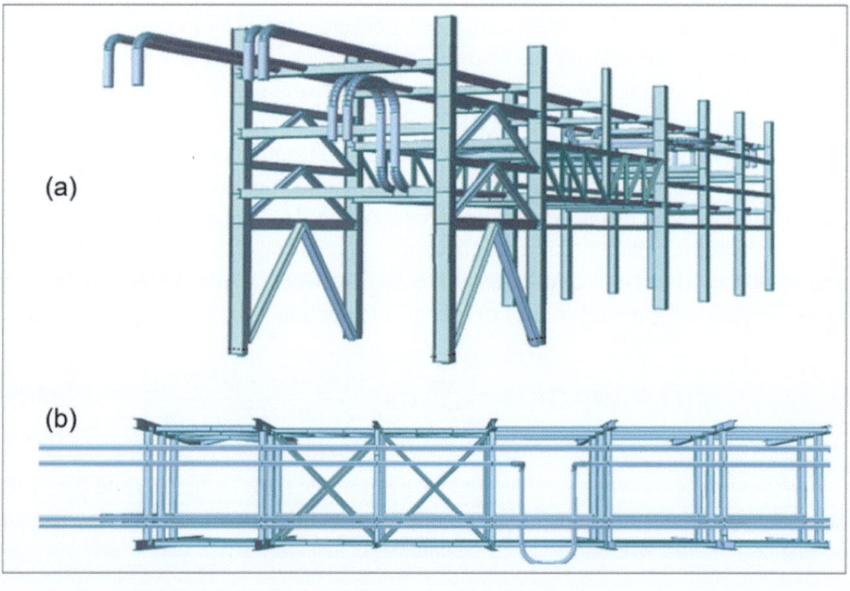

Fig. 4.10 Existing reference steel pipe-rack: **a** 3D view; **b** Top view (Randaxhe et al. 2021)

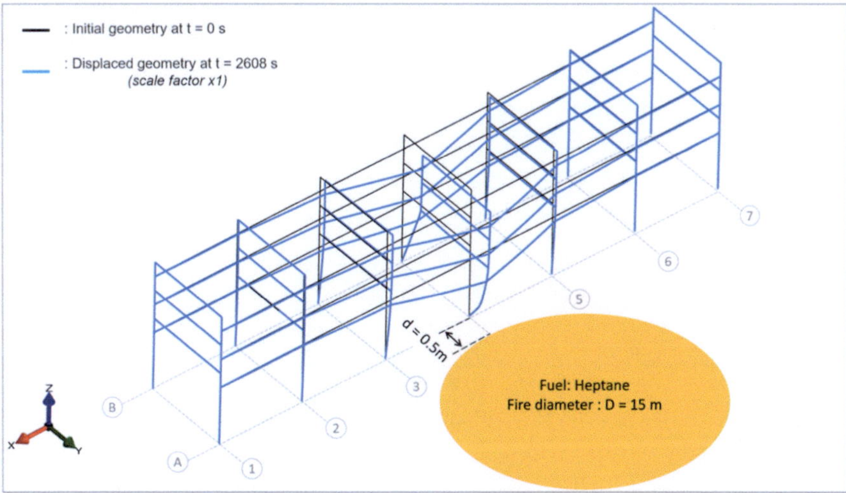

Fig. 4.11 Steel pipe-rack exposed to fire scenario "D = 15 m–d = 0.5 m – Fuel = Heptane": Failure of the structure after 2608 s (Randaxhe et al. 2021)

constant. Thus, the resulting probabilistic demand model can be represented in linear and logarithmic forms as in Eq. (4.31) and (4.32), respectively, where $a = exp(A)$ and $b = B$. A and B are parameters that can be determined from linear regression. With coefficients a and b, the standard deviation of the linear regression error can be defined with Eq. (4.33). This term is also defined $\beta_{EDP|IM}$ as the dispersion of the EDP conditioned on IM. Considering the lognormal assumption, Eq. (4.31) and Eq. (4.33), Cornell et al. (2002) defined the fragility function with the use of a lognormal cumulative distribution function as expressed in Eq. (4.34).

$$E\hat{D}P = a\ IM^b \tag{4.31}$$

$$ln(E\hat{D}P) = A + B\ ln(IM) \tag{4.32}$$

$$\sigma_{ln(EDP)|IM} = \beta_{EDP|IM} = \sqrt{\frac{\sum_{i=1}^{n}\left[ln\left(EDP_i\right) - ln\left(E\hat{D}P_i\right)\right]^2}{n-2}} \tag{4.33}$$

$$P\left(EDP > LS\middle|IM\right) = 1 - \Phi\left(\frac{\ln\left(LS/(a\ IM^b)\right)}{\beta_{EDP|IM}}\right) \tag{4.34}$$

Eventually, different EDP-IM pairs can be considered and for each of them fire fragility functions can be derived considering different LS. Based on the structural

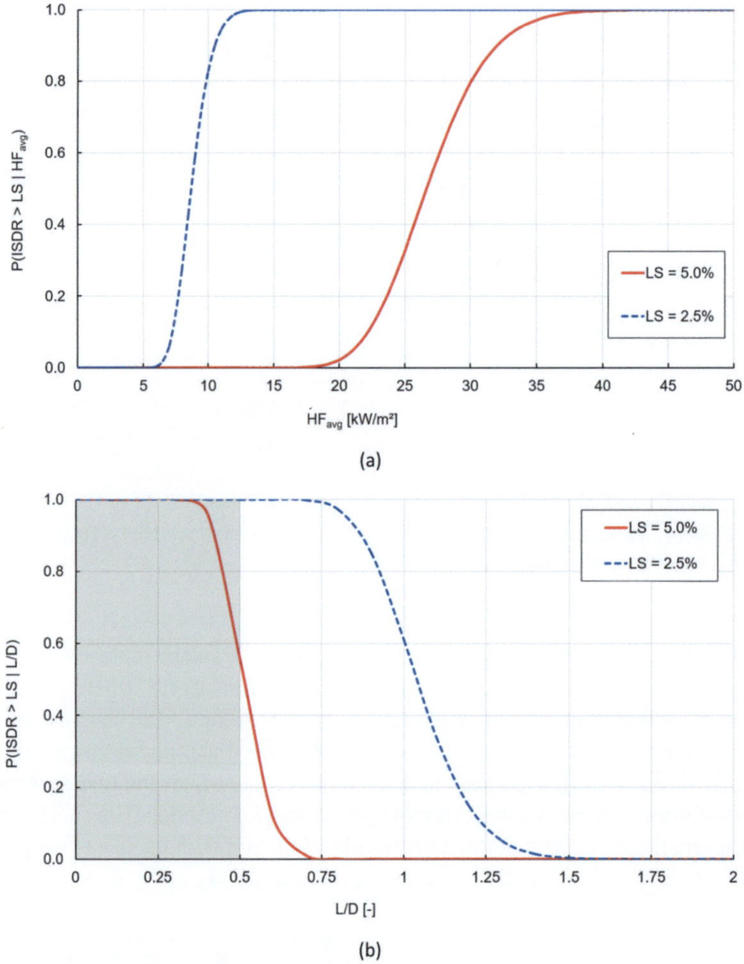

Fig. 4.12 Fragility curves based on CA for near collapse- and life safety preventions with: **a** HF_{avg} as IM and **b** L/D as IM (Randaxhe et al. 2021)

analysis outcomes, it was found that the interstorey drift ratio (ISDR)—average heat flux impinging the structure (HF_{avg}) and ISDR-ratio between fire position and fire diameter L/D EDP-IM pairs were the most efficient and was also characterised by the highest relative sufficiency among the other pairs. Finally, fragility functions considering two limit states (LS), i.e. life safety (ISDR = 2.5%) and near collapse (ISDR = 5%) limit state, were developed, as shown in Fig. 4.12 for the CA (Randaxhe et al. 2021).

4.1.10 Conclusions

In this chapter, the basis of structural fire engineering applied, with particular emphasis to industrial plants, was described. Moreover, concepts of probabilistic analysis were presented with application to a steel pipe-rack belonging to a petrochemical plant. It was shown that, in structural fire engineering, it is important:

- to define a suitable fire model, which for industrial plants in the context of the performance-based approach may be represented by localised fires;
- to perform a thermal analysis of the structural elements that is representative of the boundary conditions;
- to run a mechanical analysis that is able to capture the nonlinear response both materially and geometrically of the steel structural system. This is paramount when performing numerical analyses of part of the structure or the whole structure.

In a probabilistic fire analysis, the availability of fragility functions of a particular structural typology or component is crucial, for example, for risk assessment. This chapter presented different examples to build a probabilistic fire demand model, by adapting the CA or MSA, primarily used in the seismic field, to the fire engineering field. An example of development of a probabilistic fire demand model for a steel pipe-rack exposed to localised fires was outlined. Considering several localised plausible fire scenarios in industrial/petrochemical plants, thermo-mechanical analyses were performed with the FE method that includes the LOCAFI localised fire model. Based on numerical analyses, probabilistic analyses were conducted to derive fire fragility functions by adopting the cloud analysis (CA) method. The maximum transversal interstorey drift ratio (ISDR) was found to be the most suitable EDP for this case study, because it provides a global measure of the structural response as it can be directly related to damage and collapse of a significant part of the pipe rack. In this respect, two structural ultimate limit states, near collapse and life safety, were defined for ISDR exceeding 5% and 2.5%, respectively. The maximum average heat flux impinging the structure HF_{avg} and appeared to be the most efficient IMs to characterise localised fire severity and also the most sufficient when relatively compared with other IMs.

References

Allin Cornell, C., Jalayer, F., Hamburger, R. O., & Foutch, D. A. (2002). Probabilistic basis for 2000 SAC federal emergency management agency steel moment frame guidelines. *Journal of Structural Engineering, 128*(4), 526–533.

Anzai, K., Ban, N., Ozawa, T., & Tokonami, S. (2011). Fukushima Daiichi nuclear power plant accident: Facts, environmental contamination, possible biological effects, and countermeasures. *Journal of Clinical Biochemistry and Nutrition, 50*(1), 2–8.

Babrauskas, V. (1983). Estimating large pool fire burning rates. *Fire Technology, 19*, 251–261.

Krausmann, Elisabeth, Cozzani, Valerio, Salzano, Ernesto, & Renni, Elisabetta. (2011). Industrial accidents triggered by natural hazards: an emerging risk issue. *Natural Hazards and Earth System Sciences, 11*(3), 921–929.

Barka, A. (1999). The 17 august 1999 Izmit earthquake. *Science, 285*(5435), 1858–1859.

Bernier, C., & Padgett, J. E. (2019). Fragility and risk assessment of aboveground storage tanks subjected to concurrent surge, wave, and wind loads. *Reliability Engineering & System Safety, 191*, 106571.

Anzai, Kazunori, Ban, Nobuhiko, Ozawa, Toshihiko, & Tokonami, Shinji. (2011). Fukushima Daiichi nuclear power plant accident: Facts, environmental contamination, possible biological effects, and countermeasures. *Journal of Clinical Biochemistry and Nutrition, 50*(1), 2–8.

Elisabeth Krausmann and Ana Maria Cruz. (2013). Impact of the 11 March 2011, great east Japan earthquake and tsunami on the chemical industry. *Natural Hazards, 67*, 811–828.

Nakashima, Masayoshi, Lavan, Oren, Kurata, Masahiro, & Luo, Yunbiao. (2014). Earthquake engineering research needs in light of lessons learned from the 2011 Tohoku earthquake. *Earthquake Engineering and Engineering Vibration, 13*, 141–149.

Chang, J. I., & Lin, C.-C. (2006). A study of storage tank accidents. *Journal of Loss Prevention in the Process Industries, 19*(1), 51–59.

Committee for the Prevention of Disasters by Hazardous Materials. (2005). *Methods for the calculation of physical effects due to releases of hazardous materials–Yellow Book.* The Hague: Gevaarlijke Stoffen.

Covi, P., Tondini, N., Sarreshtehdari, A., & Elhami-Khorasani, N. (2023). Development of a novel fire following earthquake probabilistic framework applied to a steel braced frame. *Structural Safety, 105*, 102377.

Moosemiller, Mike. (2011). Development of algorithms for predicting ignition probabilities and explosion frequencies. *Journal of Loss Prevention in the Process Industries, 24*(3), 259–265.

Vílchez, J. A., Espejo, V., & Casal, J. (2011). Generic event trees and probabilities for the release of different types of hazardous materials. *Journal of Loss Prevention in the Process Industries, 24*(3), 281–287.

Ding, L. B., Pu, J. Y., & Ren, K. (2014). Analysis of thermal radiation models for large open pool fire on ship deck. *Applied Mechanics and Materials, 470*, 259–262.

Ebrahimian, H., Jalayer, F., Lucchini, A., Mollaioli, F., & Manfredi, G. (2015). Preliminary ranking of alternative scalar and vector intensity measures of ground shaking. *Bulletin of Earthquake Engineering, 13*, 2805–2840.

Elisabeth Krausmann and Ana Maria Cruz. (2013). Impact of the 11 March 2011, great east Japan earthquake and tsunami on the chemical industry. *Natural Hazards, 67*, 811–828.

Lee, Selina, Davidson, Rachel, Ohnishi, Norihito, & Scawthorn, Charles. (2008). Fire following earthquake-reviewing the state-of-the-art of modeling. *Earthquake Spectra, 24*(4), 933–967.

Covi, Patrick, Tondini, Nicola, Sarreshtehdari, Amir, & Elhami-Khorasani, Negar. (2023). Development of a novel fire following earthquake probabilistic framework applied to a steel braced frame. *Structural Safety, 105*, 102377.

Purkiss, J. (2007). Fire safety engineering. *Fire Safety Engineering Design of Structures* (2nd ed.).

Ferraz, G., Santiago, A., Rodrigues, J. P., & Barata, P. (2016). Thermal analysis of hollow steel columns exposed to localised fires. *Fire Technology, 52*, 663–681.

Menéndez, E., & Vega, L. (2010). Analysis of the behaviour of the structural concrete after the fire at the Windsor building in Madrid. *Fire and Materials: An International Journal, 34*(2), 95–107.

Mitchener, G. (2018). Impact of Grenfell tower fire disaster on polyisocyanurate industry. *Polimery, 63*.

Gardoni, P., Der Kiureghian, A., & Mosalam, K. M. (2002). Probabilistic capacity models and fragility estimates for reinforced concrete columns based on experimental observations. *Journal of Engineering Mechanics, 128*(10), 1024–1038.

Gernay, T., Khorasani, N. E., & Garlock, M. (2016). Fire fragility curves for steel buildings in a community context: A methodology. *Engineering Structures, 113*, 259–276.

Gernay, T., Khorasani, N. E., & Garlock, M. (2019). Fire fragility functions for steel frame buildings. sensitivity analysis and reliability framework. *Fire Technology, 55*, 1175–1210.

Girgin, S. (2011). The NaTech events during the 17 august 1999 Kocaeli earthquake: Aftermath and lessons learned. *Natural Hazards and Earth System Sciences, 11*(4), 1129–1140.

Godoy, L. A. (2007). Performance of storage tanks in oil facilities damaged by hurricanes Katrina and Rita. *Journal of Performance of Constructed Facilities, 21*(6), 441–449.

Peacock, R. D., Reneke, P. A., & Forney, G. P. (2015, December). *CFAST—Consolidated model of fire growth and smoke transport (version 7)*. Volume 2: User's guide.

McGrattan, Kevin, Hostikka, Simo, McDermott, Randall, Floyd, Jason, Weinschenk, Craig, & Overholt, Kristopher. (2013). Fire dynamics simulator user's guide. *NIST special publication, 1019*(6), 1–339.

Heskestad, G. (1984). Engineering relations for fire plumes. *Fire Safety Journal, 7*(1), 25–32.

Kamikawa, D., Hasemi, Y., Wakamatsu, T., & Kagiya, K. (2003). Experimental flame heat transfer correlations for a steel column adjacent to and surrounded by a pool fire. *Fire Safety Science, 7*, 989–1000.

Krausmann, E., Cozzani, V., Salzano, E., & Renni, E. (2011). Industrial accidents triggered by natural hazards: an emerging risk issue. *Natural Hazards and Earth System Sciences, 11*(3), 921–929.

Lange, D., Devaney, S., & Usmani, A. (2014). An application of the peer performance based earthquake engineering framework to structures in fire. *Engineering Structures, 66*, 100–115.

Lee, S., Davidson, R., Ohnishi, N., & Scawthorn, C. (2008). Fire following earthquake-reviewing the state-of-the-art of modeling. *Earthquake Spectra, 24*(4), 933–967.

Luco, N., & Allin Cornell, C. (2007). Structure-specific scalar intensity measures for near-source and ordinary earthquake ground motions. *Earthquake Spectra, 23*(2), 357–392.

Mudan, K. S., et al. (1984). Thermal radiation hazards from hydrocarbon pool fires. *Progress in Energy and Combustion Science, 10*(1), 59–80.

Mackie, K., & Stojadinović, B. (2003). *Seismic demands for performance-based design of bridges*. Berkeley: Pacific Earthquake Engineering Research Center.

McGrattan, K., Hostikka, S., McDermott, R., Floyd, J., Weinschenk, C., & Overholt, K. (2013). Fire dynamics simulator user's guide. *NIST special publication, 1019*(6), 1–339.

McKenna, S. T., Jones, N., Peck, G., Dickens, K., Pawelec, W., Oradei, S., Harris, S., Stec, A. A., & Richard Hull, T. (2019). Fire behaviour of modern façade materials-understanding the Grenfell tower fire. *Journal of Hazardous Materials, 368*, 115–123.

Menéndez, E., & Vega, L. (2010). Analysis of the behaviour of the structural concrete after the fire at the Windsor building in Madrid. *Fire and Materials: An International Journal, 34*(2), 95–107.

Hanus, F., Vassart, O., & Tondini, N. (2016). Nadjai, A., & Franssen, J.-M. Temperature assessment of a vertical steel member subjected to localised fire: Experimental tests. In *Structures in Fire*.

Moosemiller, M. (2011). Development of algorithms for predicting ignition probabilities and explosion frequencies. *Journal of Loss Prevention in the Process Industries, 24*(3), 259–265.

Mudan, K. S., et al. (1984). Thermal radiation hazards from hydrocarbon pool fires. *Progress in Energy and Combustion Science, 10*(1), 59–80.

Nakashima, M., Lavan, O., Kurata, M., & Luo, Y. (2014). Earthquake engineering research needs in light of lessons learned from the 2011 Tohoku earthquake. *Earthquake Engineering and Engineering Vibration, 13*, 141–149.

Possidente, L., Couto, C., Tondini, N., & Real, P. V. (2024). Review of design approaches for steel columns subjected to localised fires: Insights from the 2nd generation Eurocode model. *Engineering Structures, 315*, 118424.

Pantousa, D. (2018). Numerical study on thermal buckling of empty thin-walled steel tanks under multiple pool-fire scenarios. *Thin-Walled Structures, 131*, 577–594.

Possidente, Luca, Tondini, Nicola, & Battini, Jean-Marc. (2019). Branch-switching procedure for post-buckling analysis of thin-walled steel members at elevated temperature. *Thin-Walled Structures, 136*, 90–98.

Pantousa, Daphne. (2018). Numerical study on thermal buckling of empty thin-walled steel tanks under multiple pool-fire scenarios. *Thin-Walled Structures, 131*, 577–594.

Possidente, L., Tondini, N., & Battini, J.-M. (2019). Branch-switching procedure for post-buckling analysis of thin-walled steel members at elevated temperature. *Thin-Walled Structures, 136*, 90–98.

Possidente, L., Tondini, N., & Battini, J.-M. (2020). 3d beam element for the analysis of torsional problems of steel-structures in fire. *Journal of Structural Engineering, 146*(7), 04020125.

Possidente, L., Tondini, N., & Wickström, U. (2023). Derivation of a new temperature calculation formulation for heavily fire insulated steel cross-sections. *Fire Safety Journal, 141*, 103991.

Possidente, L., Randaxhe, J., & Tondini, N. (2023). Fire fragility curves for industrial steel pipe-racks integrating demand and capacity uncertainties. *Fire Technology, 59*(2), 713–742.

Possidente, L., Couto, C., Tondini, N., & Real, P. V. (2024). Review of design approaches for steel columns subjected to localised fires: Insights from the 2nd generation Eurocode model. *Engineering Structures, 315*, 118424.

Gardoni, P., Der Kiureghian, A., & Mosalam, K. M. (2002). Probabilistic capacity models and fragility estimates for reinforced concrete columns based on experimental observations. *Journal of Engineering Mechanics, 128*(10), 1024–1038.

Randaxhe, J., Popa, N., & Tondini, N. (2021). Probabilistic fire demand model for steel pipe-racks exposed to localised fires. *Engineering Structures, 226*, 111310.

Rew, P. J., Hulbert, W. G., & Deaves, D. M. (1997). Modelling of thermal radiation from external hydrocarbon pool fires. *Process safety and environmental protection, 75*(2), 81–89.

Shokri, M., & Beyler, C. L. (1989). Radiation from large pool fires. *Journal of Fire Protection Engineering, 1*(4), 141–149.

Shome, N., Allin Cornell, C., Bazzurro, P., & Eduardo Carballo, J. (1998). Earthquakes, records, and nonlinear responses. *Earthquake Spectra, 14*(3), 469–500.

Shrivastava, M., Abu, A. K., Dhakal, R. P., & Moss, P. J. (2019). Severity measures and stripe analysis for probabilistic structural fire engineering. *Fire Technology, 55*, 1147–1173.

Steinberg, L. J., Sengul, H., & Cruz, A. M. (2008). Natech risk and management: An assessment of the state of the art. *Natural Hazards, 46*, 143–152.

Gernay, T., Khorasani, N. E., & Garlock, M. (2016). Fire fragility curves for steel buildings in a community context: A methodology. *Engineering Structures, 113*, 259–276.

Tondini, N., & Franssen, J.-M. (2017). Analysis of experimental hydrocarbon localised fires with and without engulfed steel members. *Fire Safety Journal, 92*, 9–22.

Tondini, N., & Stojadinovic, B. (2012). Probabilistic seismic demand model for curved reinforced concrete bridges. *Bulletin of Earthquake Engineering, 10*, 1455–1479.

Tondini, N., Morbioli, A., Vassart, O., Lechêne, S., & Franssen, J.-M. (2016). An integrated modelling strategy between a CFD and an FE software: Methodology and application to compartment fires. *Journal of Structural Fire Engineering, 7*(3), 217–233.

Tondini, N., Thauvoye, C., Hanus, F., & Vassart, O. (2019). Development of an analytical model to predict the radiative heat flux to a vertical element due to a localised fire. *Fire Safety Journal, 105*, 227–243.

Gottuk, D. T., & White, D. A. (2016). Liquid fuel fires. *SFPE handbook of fire protection engineering* (pp. 2552–2590).

Committee for the Prevention of Disasters by Hazardous Materials. (2005). *Methods for the calculation of physical effects due to releases of hazardous materials—Yellow Book*. Gevaarlijke Stoffen, The Hague.

Vamvatsikos, D., & Allin Cornell, C. (2002). Incremental dynamic analysis. *Earthquake Engineering & Structural Dynamics, 31*(3), 491–514.

Vílchez, J. A., Espejo, V., & Casal, J. (2011). Generic event trees and probabilities for the release of different types of hazardous materials. *Journal of Loss Prevention in the Process Industries, 24*(3), 281–287.

Randaxhe, Jérôme., Popa, Nicoleta, & Tondini, Nicola. (2021). Probabilistic fire demand model for steel pipe-racks exposed to localised fires. *Engineering Structures, 226*, 111310.

Yang, J.-E. (2014). Fukushima dai-ichi accident: Lessons learned and future actions from the risk perspectives. *Nuclear Engineering and Technology, 46*(1), 27–38.
Zheng, B., & Chen, G. (2011). Storage tank fire accidents. *Process Safety Progress, 30*(3), 291–293.

Chapter 5
A Probabilistic Approach for Seismic Resilience Analysis of Hazardous Industrial Facilities in Seismic Areas

Fabrizio Paolacci, Antonio Caputo, Daniele Corritore, and Gianluca Quinci

Abstract Industrial plants can be highly susceptible to seismic activity. Historical seismic events have highlighted the severe impact and significant economic losses that an industrial plant can suffer, not only due to physical damage to equipment but also due to disruptions in production processes. To quantify these economic losses, an evaluation of the plant's seismic resilience is necessary. This paper introduces a general probabilistic framework for assessing industrial plant resilience and economic losses in the event of seismic activity. The framework considers uncertainties related to the plant equipment's ability to withstand disturbances and the recovery process, including equipment recovery times and costs. Generated samples by Monte Carlo simulations are employed to derive these uncertainties. A black carbon plant serves as a case study to demonstrate the model's applicability. The suitability of the proposed model indicate that it represents a valuable tool for decision-makers, plant owners, insurance companies, emergency managers, and plant designers.

5.1 Introduction

Industrial facilities are recognized to be highly susceptible to the effects of natural disasters, catastrophic failure of equipment, or supply chain interruption. The related consequences can rapidly propagate through the supply chains, with consequences on the production capacity and with important business interruption effects. Natech event, acronym of Technological accidents triggered by natural hazards, appears particularly critical because of the severe effects that can be generated. In general, damage to industrial facilities generated by earthquake may severely

F. Paolacci (✉) · G. Quinci
DICITA, Roma Tre University, Via vito volterra 62, Rome, Italy
e-mail: fabrizio.paolacci@uniroma3.it

A. Caputo
DIIEM, Roma Tre University, Via vito volterra 62, Rome, Italy

D. Corritore
SAFEPLANT srl, P. le R. Ardigò 20/A, Rome, Italy

V. Cozzani and F. Paolacci (eds.), *Management Strategies and Resilience Against NaTech Events*, CISM International Centre for Mechanical Sciences 616,
https://doi.org/10.1007/978-3-031-84485-0_5

impact as demonstrated in past seismic events (Cruz and Steinberg 2005; Mesa-Gomez 2020). In particular, hazardous industrial facilities as oil refineries shown a high vulnerability to seismic loading, like for example during the 2011 Tohoku earthquake (Kajitani et al. 2013). In such a cases risk-based assessment measures, which focus on vulnerability and immediate consequences are not sufficient, while resilience is a performance measure providing a more complete picture. Resilience measures both the robustness of a system, i.e. the ability to survive unexpected disruptive events, and the capability of rapidly restoring its capacity after the event occurred (Bhamra 2011; Bristow and Hay 2017; Hosseini et al. 2016a; Ivanov et al. 2017; Woods 2015). Concerning the system capabilities of absorption, adaptation and restoration, resilience is of great relevance to operations managers, plant engineers, public administrators, and industrial stakeholders in general. Reports on several examples of the devastating consequences of lack of resilience in industrial sector are reported in Sheffi (2005; 2015). Thus, predict unexpected disruptions, by computing resilience-based metrics, is a priority for industrial companies, Nauck et al. (2021), and is a key ingredient for effectively mitigating adverse events impact through creating more resilient systems. Methods for quantifying the resilience of industrial systems began more than two decades ago, and were initially aimed at civil infrastructure, networked systems, and utility distribution infrastructure (e.g., telecommunication and road transportation networks, gas and electricity distribution networks or telecommunication networks, etc.), as well as at the utility distribution infrastructure, Argyroudis and Mitoulis (2021), Sun et al. 2019). The extensive economic damage caused by earthquakes to industrial facilities recorded during recent seismic events has shifted the focus to these critical infrastructures, although there has been relatively little research on estimating the resilience of industrial facilities, including process plants, El-Halwag (2020). In this regard, the literature shows that two main approaches can be used for resilience modeling: Flow-based approaches and non-flow-based approaches. In particular, non-flow-based methods focus on the logical interaction between equipment while neglecting the physical structure of the plant in many cases, thus making the estimation of process flows more difficult if not impossible, especially if different process flows coexist. They also do not allow for the calculation of recovery time and capacity because they do not take into account the interactions between recovery activities, with specific end-start constraints. Typically, the recovery process is often modeled using predefined analytical functions (Cimellaro et al. 2009; Cimellaro 2006) that are often unrealistic. For example, the use of continuous instead of discrete recovery curves does not represent the real structure of the process nor does it include the recovery actions suggested by the logical constraints of the recovery tasks. Concerning non-flow-based models, the main adoptable approaches are Bayesian networks (BNs) or reliability-based modeling. Typically, BNs do not describe the physical structure of the system, but reproduces the logical "cause-consequences" relations between the damage caused by the event and the consequences. In this respect probabilistic aggregate indicator of system resilience are provided, and evaluating this indicator over different time intervals allows the evolution of resilience over time to be tracked instead of a temporal trend of capacity based on the progress of actual recovery activities. Non-flow-based models were applied to the case of the Fukushima Daiichi nuclear power plant accident (Abimbola and Khan 2019) the Chevron refinery accident (Tong et al. 2020), a sulfuric As typical non-flow-based model, As typical non-flow-based model, acid production plant (Hosseini et al. 2016b). In contrast, flow-based approaches are based on modeling the physical structure of systems and computing one or more types of flows through the system. This type of approach is therefore intrinsically appropriate to modeling process systems. Flow-based models duplicate system structure by mapping the physical and logical interconnections between system components. Such models can be supported by mathematical programming methods in which the value of decision variables must be selected to optimize a performance measure during system design or recovery after an outage. Regarding the seismic resilience of industrial plants, epistemic uncertainty or randomness should also be included. In this respect, Caputo and Paolacci (2017) developed a method based on process flows, functional block diagrams and activity networks to model both the impact of outages on system capacity and the discontinuous time course of capacity recovery dictated by the actual interaction between equipment restoration activities. The model is imagined for process plants under natural hazards and allows to compute economic loss in terms of reconstruction and business interruption costs; in this sense is particular functional to the expectations of risk managers. Mussini (2019) develops a similar method, where, however, the interaction between recovery actions is only captured in approximate manner. Consequently, to model the resilience of process plants, a flow-based approach seems to be more efficient, especially in case of natural events, like earthquakes, because it allows the calculation of vulnerability and resilience, based on the actual physical structure of the plant, taking also into account the effect that equipment failures have on changes in process flows, with the impact on production loss costs. Furthermore, in order to reliably calculate the length of the recovery period, it is necessary to take into account the mutual interactions between the recovery activities of individual units. The ability to calculate the actual time course of capacity recovery, also allows for the calculation of various resilience metrics and easy assessment of the contribution of individual equipment to plant resilience. However, a comprehensive probabilistic approach that fully implements the capabilities of a flow-based approach is currently lacking. For these reasons, a probabilistic flow-based model is proposed in this chapter for estimating the resilience of process equipment in case of seismic Natech events. It is based on a simplified graph network representing the recovery activities, showing the interdependencies between the reconstruction activities of different equipment. At this end, a specific probabilistic model of equipment recovery phases is formulated. The method has also the ability to estimate the economic losses resulting from repair activities, material loss and business interruption, along with the definition of the recovery tasks, which is useful for project management purposes. A specific seismic hazard analysis module allows to generate multiple damage scenarios for Natech seismic events. This enables facility designers and safety managers to explore the possible impact of high-risk, low-probability events and plan effective mitigation strategies to increase the resilience of industrial facilities during earthquakes. The chapter is organized as follows. Section 5.2 deals with a general formulation of probabilistic resilience analysis of process plants (PSRA), and a simplified formulation of the PSRA model for process plants that are dedicated to the transformation of raw materials in final products, and for this reason particularly diffused. An illustrative example of seismic resilience quantification is analyzed in Sect. 5.3, where a black carbon plant under earthquakes is analyzed to show the practical application of the methodology and its capabilities. Discussion of results and prospects for future research conclude the paper.

5.2 Probabilistic Seismic Resilience Analysis (PRA) of Process Plants

5.2.1 General Formulation of Natech Resilience for Industrial Plants

Seismic resilience of a process plant can be defined as the capability of a plant to survive a low frequency high impact disruptive earthquake and recover to maximize the total production capacity in case of physical damage. Plant operational capacity C(t), which represents the physical production output of the plant, can be used to calculate the resilience of the system. In Fig. 5.1 a schematic of the time trend of C(t) is synthetically illustrated, which is characterized by five different states:

a. the pre-earthquake state which is defined as the nominal plant capacity C(t0), which can be increased by upgrades made by stakeholders, or reduced by aging effects (Ayyub 2014). Propagation of damage within the plant caused by loss of containment effects, known as domino effects (Antonioni et al. 2009c; Alessandri et al. 2018; Paolacci et al. 2018; Chen et al. 2020), impose the condition $t0 < td$. Finally is it important to stress the fact that in case of domino effects the weather conditions and plant emergency response will play an important role in the estimation of the plant residual operational capacity C(td);
b. The post-earthquake condition, from, td to ta, corresponds to the inspection and planning phase necessary to identify all damage and decide on the recovery plan. The extent of this state is mainly influenced by the preparedness of the facility's emergency managers and the existence of a recovery plan. In addition, the resilience of other systems, such as critical infrastructure or transportation infrastructure, may have some influence (Wu et al. 2020; Unnikrishnan et al. 2016; Gardoni et al. 2019, 2021). Delays due to obstructive factors (post-earthquake inspection, engineering, permitting, financing, mobilization of contractors) may affect the duration of this phase, which ends at a time ta, the start time of the reconstruction phase.
c. The recovery process, in which the recovery path is usually represented by a step function from ta until the complete recovery of the operational capacity, (tr). In fact, the recovery process sequentially involves the restoration of equipment,

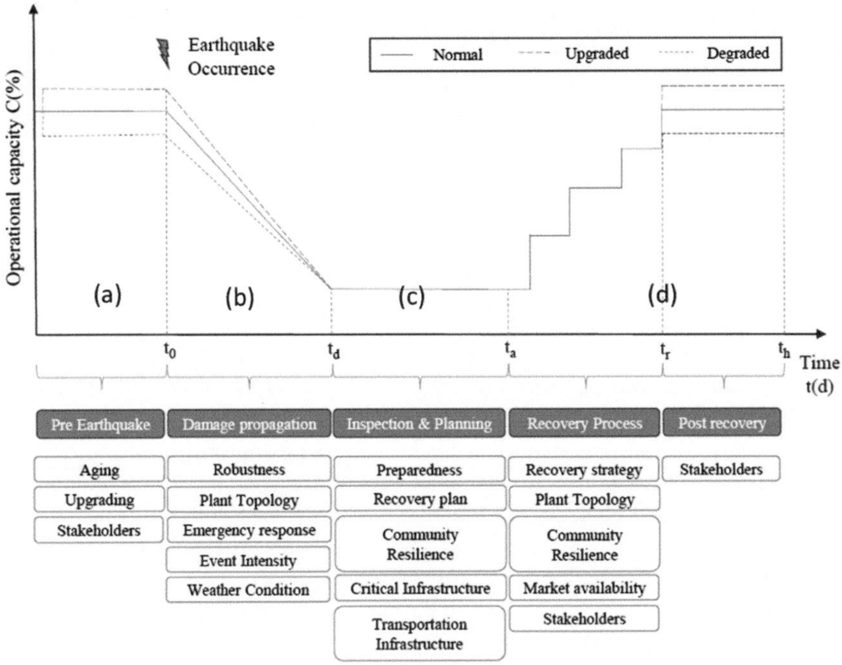

Fig. 5.1 Plant operational capacity versus time (Kalemi et al. 2024)

whose capacity is considered instantaneously recovered generating a step path. See also Mussini (2019).

d. The last state, corresponding to the post-recovery steady state, has a duration in the interval th-tr, where th is a control time, generally imposed by the stakeholders (Cimellaro et al. 2009).

Different indexes can be adopted for the quantification of resilience R (Castillo-Borja 2018; Jain 2018). In the present paper, R is calculated as the integral over time of the operational capacity curve, using the proposal of Cimellaro et al. (2009), according to Eq. (5.1). The reason is mainly related to the dimensionless form of R and its smooth nature. A plant resilience index can be adopted for the quantification of resilience (R) and calculated in different manners, as already mentioned in the introduction. In the present paper R is calculated as the integral over the time of the operational capacity curve, using one of the most popular expression present in the literature (Cimellaro et al. 2009), according to Eq. (5.1). The reasons of this choice are basically related to the non-dimensional form of R and its smooth nature related to the integral operation adopted.

$$R = \frac{1}{t_h - t_0} \int_{t_h}^{t_0} C(t)\,dt \tag{5.1}$$

However, other Resilience indexes could use as alternative to Eq. (5.1), including residual operational capacity $C(td)$, time to full capacity recovery (tr), maximum recovery interval $(Tmax = tr - td)$ and also total economic losses (EL) which includes direct and indirect costs, according to Federal Emergency Management Agency (2012). Their analysis will be leave to future research.

5.2.2 *Probabilistic Seismic Resilience Analysis (PRA)*

General Formulation. In order to develop a full probabilistic approach in resilience quantification, and taking into account the variability of all parameters involved in the calculation a full probabilistic approach has been proposed by Kalemi et al (2024) and here summarized. The idea is to follow the classical PBEE approach used for the quantification of seismic risk. Beyond the uncertainties in seismic hazard and equipment vulnerability (Bakalis et al. 2018; Karaferis et al. 2022; Kazantzi et al. 2022; Melissianos 2022; Phan et al. 2017), the uncertainties in recovery should affect duration of equipment/activities recovery due to work delays, weather delays, supply chain delays, etc. In addition, uncertainties in the cost of recovering equipment/activities due to unexpected construction work or due to variable costs of equipment parts and raw materials should also be considered in the quantification of recovery interval (TR). Consequently, the general formulation of probabilistic resilience R, including all uncertainties mentioned above, can be expressed as:

$$
\begin{aligned}
E[R(t)] = \nu \int \cdots \int r(t)\, f(R(t)|TR)\, f(TR|DM)\, f(DM|EDP)\, f(EDP|IM) \\
f(IM|d, m) \times f(d)\, f(m)\, dR\, dTR\, dDM\, dEDP\, dIM\, dD\, dM
\end{aligned}
\tag{5.2}
$$

where $E[R(t)]$ is the expected annual resilience index $R(t)$, expressed as function of the random variables and time t elapsed from $t0$; ν is annual rate of occurrence of a seismic event with magnitude $M > Mmin$, provided by the Gutenberg-Richter law. The letter f indicates the probability density function (PDF) of the random variables: recovery interval (*TR*), damage Measure (*DM*), engineering demand parameter (*EDP*), seismic intensity measure (*IM*), Magnitude (M) and distance from the fault (D). In case of different seismic sources, the result can be simply determined as summation of the single contributions. The expected annual economic losses can be quantified in a similar manner considering the economic loss function L(t) instead of the resilience R(t) and where the recovery cost C is considered a random variable as well. A direct way to solve Eq. (5.1) is to use a Monte Carlo Simulation (MCS) approach. For each simulation, samples of Magnitude M, Distance D, Seismic Intensity IM, Apparatus damage DM, Recovery Time TR and recovery cost C and derived by their respective Pdf functions f. The advantages of this approach are clear: (a) the expected value of resilience E[R(t)] and mean annual losses E[L(t)] can also be easily evaluated, (b) the formulation is rather general which allows the calculation in a unique framework of the statistic of all random variables of interest. Nevertheless, the computational effort is generally extremely high and could be prohibitive. A different approach, already proposed by the authors in Kalemi et al. (2024) is to evaluate the statistics of L and R in a discrete manner for different values of IM (seismic scenarios), and sampling values of recovery time, cost and damage with a MCS approach. Consequently, it is possible to associate a frequency of occurrence to each IM through a classic PSHA (Cornell 1968; Baker 2008) and thus evaluate the mean annual frequency of economic losses and resilience metrics. Obviously, the quality of the method depends strictly on the number of selected PGA values.

Preparatory Steps for MCS. The implementation of the proposed PRA method needs a series of preparatory sub-steps in order to allow running MCS and thus derive, for each scenario (PGA) samples of Operational Capacity Curve, Resilience R(t) and Economic losses L(t). In particular,

- Process mapping and plant topology representation
- Definition of initial residual capacity of the plant
- Formulation of plant recovery model
- Formulation of the plant recovery function
- Definition of resilience index and economic loss model
- Probabilistic Seismic Hazard Analysis and evaluation of seismic damage scenarios.

Firstly, the process map must be identified and plant topology described. Generally, the process maps can be effectively represented by Process Flow Diagram (PFD) of the plant, which includes the main equipment of the plant and mapping the different material flows from one equipment to others. An equipment can belong a unique process flow (PF) or multiple process flows contemporary. In order to represent all the equipment that are included in the f-th process flow, a unique set of equipment S[f] should be defined (Caputo et al. 2020). For each set of equipment S a Capacity Block Diagram (CBD) can be constructed (Caputo et al. 2020). CBD contains process stages (PS) connected in series, where the capacity of the entire process flow (PF) corresponds to the capacity of the PS with the lowest capacity. A process stage contains a group of equipment collected based on the operational functionality and influence on the operational capacity of the PF. Process stages could contain equipment in series or in parallel (redundant). The practical application of the CBD can be clarified from the case study presented in the next section. According to a seismic scenario generated by selecting a seismic intensity measure IM, it is possible to account for the damage condition of the equipment of each PF by a vector DSV (Damage State Vector) containing a code related to the damage state in which each equipment is (1-no damage 2-DS = 2, etc.). In the following the PGA will be used as IM (Caputo et al. 2020). Consequently, for each recovery phase the damage condition of each equipment is monitored and updated though a State Vector SV that contains a number of elements equal to the number of equipment, which are 0 if the equipment is damages or 1 if is not damaged or repaired. Using this representation, the residual overall plant capacity C(td) can be easy evaluated as fraction of the initial Capacity (t0), simply updating the status of the equipment (undamaged, Damaged, Repaired). A strategic point of the proposed methodology is the formulation of plant recovery model. Different recovery models can be implemented for the recovery process of industrial plants (Fig. 5.3). The Overall Reconstruction Activity Network (ORAN) model, was introduced by Caputo and Paolacci (2017) and was applied by Kalemi et al. (2019) and Caputo et al. (2020). The ORAN model allows to describe the recovery process with an activities network and is based on CPM/PERT method (Vanhoucke 2012) for calculation of the time of the equipment recovering (TCRi). The ORAN model is very detailed and accurate as it takes into account the interdependencies between recovery activities. However, the main drawback is the necessity of a great number of information to be applied, with the consequence that often results no practical, especially when a quick evaluation of risk and resilience analysis is required. Consequently, different levels of simplification shall be implemented, as illustrated in Fig. 5.3. In this work a simple approach based on a simplified representation of plant equipment recovery network is used, similarly to Almufti and Willford (2013). In particular, the inspection and planning recovery task is considered unique, whereas the equipment recovery tasks start just after contemporarily (Fig. 5.3c) or in series if an interdependency is present (Fig. 5.3d). With this approach, the recovery process is considered, for each damaged equipment, as a single task whose duration is related to the damage state. The duration of equipment recovery can be calculated based on different tasks (engineering, permitting, delivery and installation), the duration of which can be assumed from literature (e.g. FEMA 2022) or can be defined based on expert judgement. Consequently, the probability distribution functions of each duration of the inspection, design phase and the equipment recovery phase can be defined for the implementation of the Monte Carlo simulations. The main drawback is the assumption of interdependency between equipment that can be assumed either at the start of equipment recovery process or when the equipment is fully recovered. Moreover, it does not account for any possible interconnections during the recovery process. At the end, the method will allow the calculation for each equipment the recovery time TCR. More details on the method can be found in Kalemi et al. (2024). For each seismic damage scenario, the resilience R is quantified according to Eq. (5.1), while the expected losses (EL) will be calculated as the sum of direct (DC) and Business Interruption (BI) cost. DC are determined according to Eq. (5.3) as sum of cost of preparatory tasks (CP) and the ith equipment reconstruction costs (RCi).

$$DC = CP + \sum_i RC_i \tag{5.3}$$

$$BI = \sum_f \sum_z (p_f - Cvu_f)[Cn_f - C_f(t)]\Delta t_z \tag{5.4}$$

In Eq. (5.4), with reference to the fth process flow, Cf(t) is the capacity at time t, Cnf is the nominal production output, Cvuf is the variable unit production cost, pf is the unit vending price, whereas Δt_z is the duration of the zth time interval between

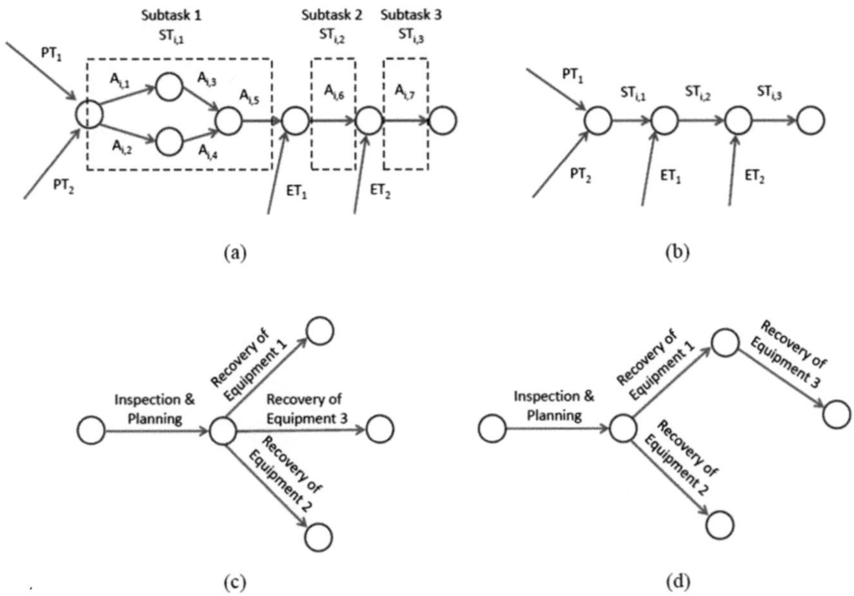

Fig. 5.2 **a** Example of equipment subtask grouping of a fictitious equipment; **b** Simplified activity network of a fictitious equipment; **c** SPEREN: all equipment start recovering instantaneously; **d** SPEREN: some equipment start recovering after the recovery of the equipment they depend on. Legend: A = reconstruction activity, PT = preparatory task, ET = external task, ST = subtask

functional recovery of two successive units (Caputo et al. 2020). The proposed simplified methodology relies on different damage scenarios to be used in the Monte Carlo framework of Fig. 5.2. In this respect the following steps are deleveloped:

- Probabilistic Seismic Hazard Analysis (PSHA) of the plant site;
- Vulnerability analysis of all equipment;
- Evaluation of seismic damage scenarios of the plant;

These aspects have been largely treated in Alessandri et al. (2018), where a new method for quantitative seismic risk analysis of major-hazard process plants has been proposed. In order to generate seismic damage patterns of the plant equipment, the PRIAMUS software proposed in Corritore et al. (2017) is used. This software, based on a Probabilistic Seismic Hazard analysis of the site and Fragility curve of each equipment, easily generates damage patterns and damage propagation effects (domino effects). The method uses Monte Carlo simulations to generate samples of damage scenarios involving the equipment of a plant exposed to specific seismic hazard conditions, based on which the relevant statistics are constructed.

5.2.3 Quantification of Probabilistic Resilience Metrics Through Monte Carlo Simulations

The framework illustrated in Fig. 5.3 synthetizes the proposed method for Resilience quantification of a process plant in terms of Economic Losses and Resilience Index. For each seismic damage scenario corresponding to step 1, which is characterised by a PGA value, a Monte Carlo Simulation (MCS) is performed to quantity Resilience Index through Eq. (5.2) or in a similar manner the Economic Losses. Given that now the seismic damage scenario is known, the integral of Eq. (5.2) can be simplified as follows:

$$E[R(t)|PGA_j] = \int_R \int_{\mathbf{T_{CR}}} \int_{\mathbf{DSV}} r(t) \times f_{(R|\mathbf{T}_{CR})} f_{(\mathbf{T}_{CR}|\mathbf{DSV})} f_{(\mathbf{DSV}|PGA)} \, dR \, d\mathbf{T}_{CR} \, d\mathbf{DSV} \tag{5.5}$$

where now the independent random variables are the damage scenario vector DSV, the recovery time vector TCR. Similarly business interruption losses BI(t) and direct costs DC can be derived using the same integral form, substituting r(t) with bi(t) or dc. In brief, for each MCS step and for each equipment, a random sampling of the damage scenario vector DSV along with corresponding equipment state vector SV are derived through fragility curves, as shown in Fig. 5.3, Alessandri et al. (2018). Consequently, the recovery dates vector of each equipment (TCR) can be evaluated summing the time for inspection planning phase Tp, common to all equipment, the time of the ith activity TA(i) and the maximum recovery activity of the previous activities (max

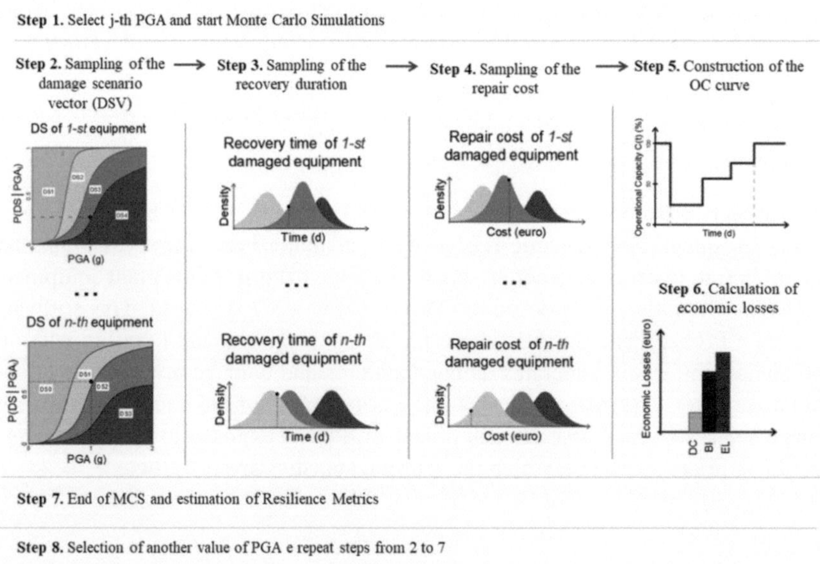

Fig. 5.3 The framework of the proposed PRA method

recovery time of the predecessors). The logical sequence of the method is shown in Fig. 5.3. The Monte Carlo simulations are repeated NPGA times, and statistics on resilience metrics and economic losses (direct and indirect) can be built for each scenario. The mean annual losses (EALm) can be finally calculated as integral of the loss exceedance function versus the mean annual frequency of exceedance PGA, λ. In this respect, given the limited number of earthquake scenarios, EAL can be approximated by Eq. (5.6). The same procedure can be used to separately calculate business interruption BI and direct costs DC.

$$EAL_m \approx \sum_{j=2}^{N_{PGA}} \frac{EL_m(PGA_j) - EL_m(PGA_{j-1})}{2}[\lambda(PGA_j) - \lambda(PGA_{j-1})]. \quad (5.6)$$

5.3 Illustrative Example

5.3.1 Description of the Case Study

A black carbon plant is selected as illustrative (Karagiannakis et al. 2020). It is assumed to be ideally located in Priolo Gargallo (Italy). The plant produces different types of black carbon, which are used for coloring products in black tones, or as additives in various rubber products, such as car tires, car dashboard etc. A plan view of the plant is shown in Fig. 5.4. The principal process is the following: (i) the feed stock oil is stored in the fuel tanks; (ii) the oil is preheated and driven to the reactors (oil pumps, pipe rack, heat exchangers); (iii) an incomplete combustion of the oil performed in the reactors produces carbon black powder, and tail gas (horizontal and vertical reactors); (iv) the carbon black powder is separated from the gas and compacted into pellets (milling towers); finally, the pellets are stored in silos before they will be sold. Twenty-three major equipment are included in analysis as shown in Fig. 5.5. In the plant two process flows (PFs) are identified, corresponding to two production lines sharing the pumps, and it is capable to produce two different types of black carbon contemporarily. PF1 has a production capacity of 60 t/d while PF2 has a production capacity of 100 t/d. Both types of carbon black that are analyzed in this case study are characterized by a unit production cost of around 1110 €/t and a market price of about 1270 €/t, based on information provided from plant engineers (Kalemi et al. 2024). In the same Fig. 5.5 the CDB diagram of the plant is shown, where series or redundant, as well as fractional equipment are reported. The plant is assumed to be located at a site with a latitude of 39.17° and a longitude of 15.17°, characterised by a soil type B. The seismic hazard curve of the selected site reported in Alessandri et al. (2018). The PGA valued used in the analysis range between 0.05 and 1.79 g. The maximum value of PGA considered in the analysis, i.e. 1.79 g, corresponds to a very low annual probability of exceedance, which is about 10E-6. Seismic vulnerability of equipment is generally represented by fragility curves. Equipment summarized in Fig. 5.6 have multiple damage states. The definition of damage states of equipment

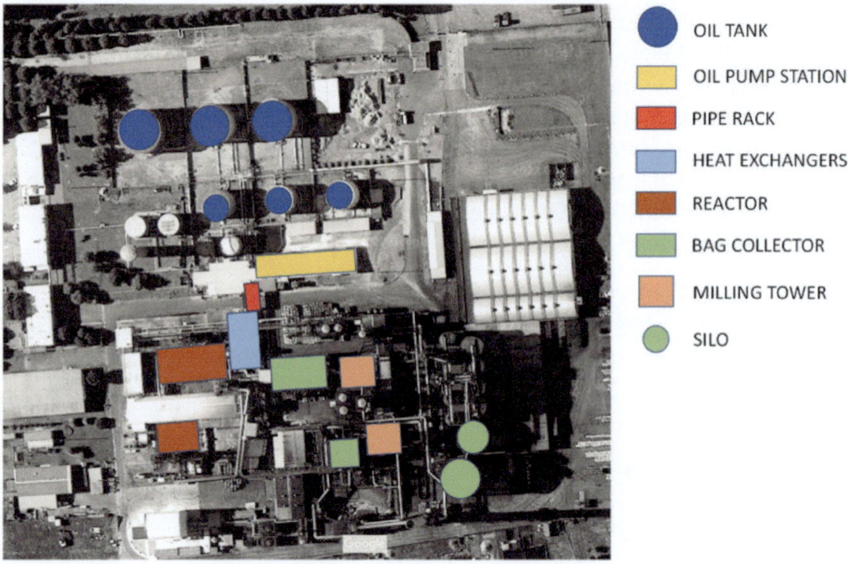

Fig. 5.4 Plan view of the black carbon plant

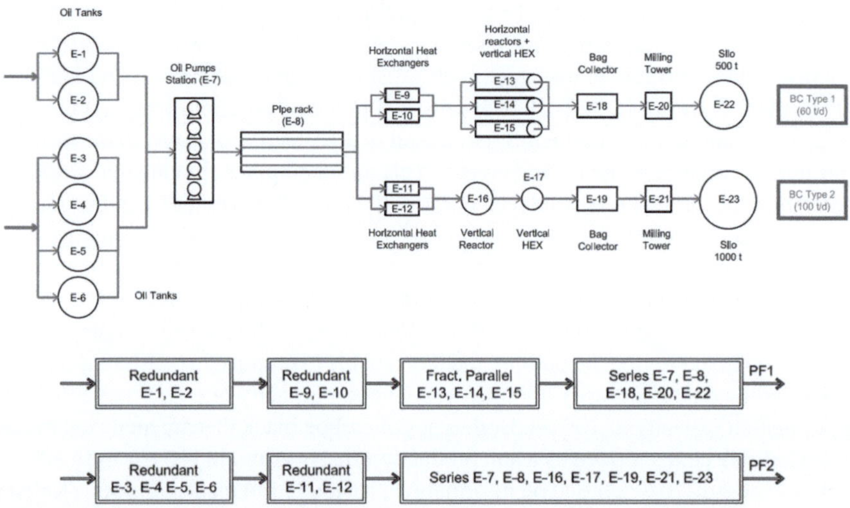

Fig. 5.5 Simplified process flow diagram of black carbon plant and CBD diagram

Equipment #	Type of Equipment	DS2		DS3		DS4		DS5	
		PGA_m (g)	β	PGA_m (g)	β	PGA_m (g)	β	PGA_m (g)	β
E-1, E-2, E-3, E-4, E-5, E-6	Steel storage tank	0.15	0.70	0.35	0.75	0.68	0.75	0.95	0.70
E-7	Oil pumps station	0.15	0.75	0.34	0.65	0.77	0.65	1.50	0.80
E-8	Pipe rack	0.16	0.79	0.35	0.69	0.49	0.72	-	-
E-9, E-10, E-11, E-12	Horizontal HEX	-	-	0.52	0.70	-	-	1.56	0.61
E-13, E-14, E-15	Reactor + Vertical HEX	-	-	0.60	0.40	0.88	0.39	-	-
E-16	Vertical reactor	-	-	-	-	0.97	0.71	-	-
E-17	Vertical HEX	-	-	0.60	0.40	0.88	0.39	-	-
E-18, E-19	Bag collector	0.22	0.71	3.05	0.71	5.62	0.71	-	-
E-20, E-21	Milling tower	0.36	0.64	0.76	0.64	1.80	0.64	-	-
E-22	Silo 500t	0.74	0.77	1.26	0.77	7.56	0.77	-	-
E-23	Silo 1000t	0.74	0.77	1.26	0.77	7.56	0.77	-	-

Fig. 5.6 Fragility curves parameters for different damage state of equipment (after Kalemi et al 2024)

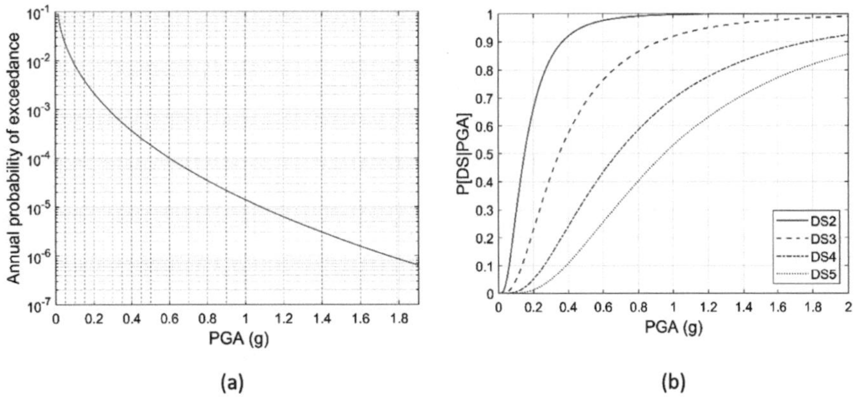

Fig. 5.7 **a** Seismic hazard curve of Priolo Gargallo site with the selected PGA; **b** Hazus (2022) Fragility curves of oil tanks

is adopted from FEMA (FEMA 2022), where DS1 corresponds to no damage, DS2 corresponds to slight/minor damages, DS3 corresponds to moderate damage, DS4 corresponds to extensive damage, and DS5 to complete damage. For more details about seismic vulnerability and damage states of the main equipment of hazardous facilities, the readers can refer to Paolacciet al. (2013). Figure 5.7 shows the seismic hazard curve of the selected site and the fragility curve for oil tanks (FEMA 2022).

5.3.2 Simplified Recovery Model

The proposed Simplified Probabilistic Equipment Recovery Model is used and the equipment recovery duration and costs are reported in Fig. 5.8. Mean values of equipment recovery duration and cost are indicated in Karagiannakis et al. (2020). In this study, both recovery times and costs of equipment are assumed as truncated normally distributed. A coefficient of variation 0.2 is assumed for both recovery times and repair costs. It is also assumed that the reconstruction of all damaged equipment start at the same time, after the inspection and planning activities, and without interdependencies. A control period of two years (th = 730d) is chosen to analyse the operational capacity curve of the plant and to calculate the seismic resilience index. For each PGA value, a Monte Carlo simulation has been performed using the PRIAMUS software (Corritore et al. 2017) to generate a random seismic damage scenario. For each equipment damage states DS, a random recovery duration and a random recovery cost are sampled from their distributions. An example of seismic damage probability of exceeding multiple damage states in the equipment for PGA = 0.5g is shown in Figs. 5.9.

5.3.3 Description of Results and Discussion

Figure 5.10a shows the operating capacity curves of the black carbon plant for different confidence intervals (from 10th to 90th), for a PGA = 0.5g. It can be noted as the 90th percentile operating capacity curve does not reach full capacity even after 730

Process Equipment	DS2		DS3		DS4		DS5	
	Cost (€)	Recovery Time (d)	Cost (€)	Recovery Time (d)	Cost (€)	Recovery Time (d)	Cost (€)	Recovery Time (d)
Steel storage tank	25,000	15	250,000	210	500,000	300	600,000	360
Oil pumps station	750	3	7,500	10	15,000	90	16,500	100
Pipe rack	16,600	10	58,100	70	83,000	135	-	-
Horizontal HEX	-	-	65,000	185	-	-	78,000	225
Reactor + Vertical HEX	-	-	1,050,000	120	2,100,000	660	-	-
Vertical reactor	-	-	-	-	3,000,000	660	-	-
Vertical heat exchanger	-	-	365,000	60	730,000	295	-	-
Bag collector	60,000	15	600,000	120	1,200,000	390	-	-
Milling tower	35,000	15	350,000	120	700,000	660	-	-
Silo 500t	25,000	10	250,000	90	500,000	480	-	-
Silo 1000t	35,000	10	350,000	90	700,000	480	-	-

Fig. 5.8 Mean values of recovery time and recovery cost for each DS of equipment (after Kalemi et al. 2024)

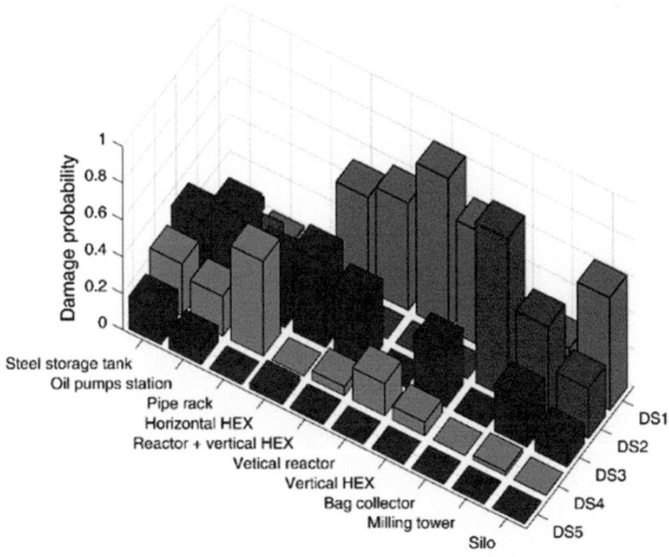

Fig. 5.9 Seismic damage Probability of exceeding damage states (DS) in the equipment for PGA = 0.5g

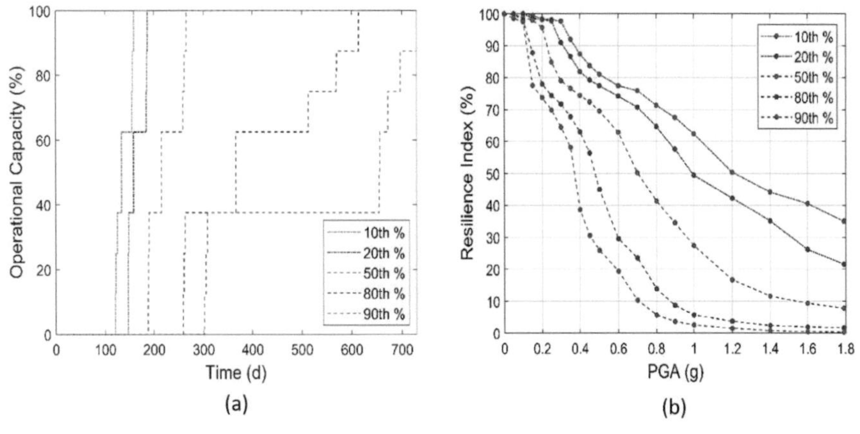

Fig. 5.10 **a** Probabilistic operational capacity curves of black carbon plant for MCS corresponding to PGA = 0.5; **b** Probabilistic resilience index curves for different PGA

d. This is evidently due to delays resulting from recovery of damage scenarios with at least one vertical reactor, reactor and vertical heat exchanger, or milling tower in DS4, which have a mean recovery time of 660 d (see Table 5.8). Figure 5.10b illustrates the resilience index for different PGA levels. It can be noticed that for PGA < 0.15 g the resilience indexes does not vary excessively; differently, for PGA ≥ 0.15 g the variation is evident. Concerning the 80th and 90th percentiles resilience index, it

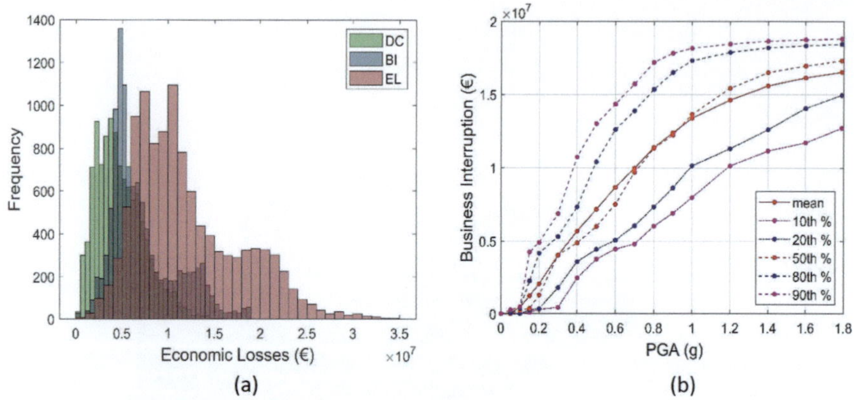

Fig. 5.11 **a** Distribution of economic losses for PGA = 0.5 g; **b** Business interruption losses for different PGA

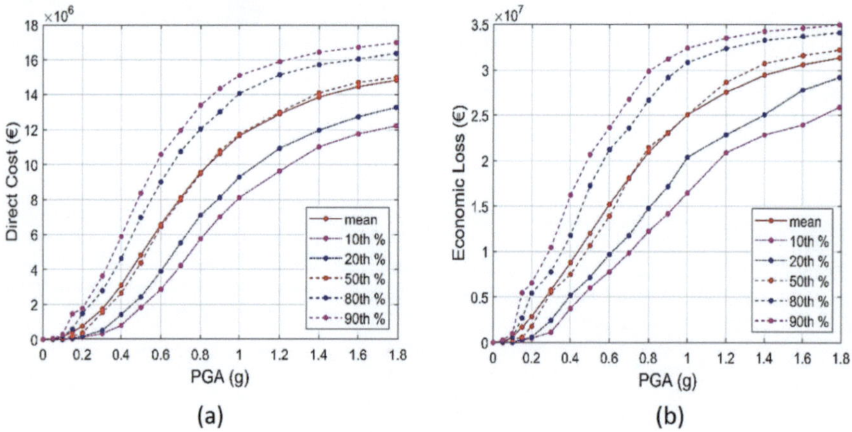

Fig. 5.12 **a** Direct cost losses for different PGA; **b** Economic losses for different PGA

can be noticed a difference of 10–20% in the range of PGA from 0.35 to 0.75 g while for other values of the resilience the difference become smaller. Moreover, for PGA $\geq$1 g the 80th and 90th resilience index drops almost to 0%. Figure 5.11a shows the probability distribution function of the economic losses, for PGA = 0.5 g. Direct costs presents a mean value of 4.81 million euros, BI have a mean 7.17 million euros and EL have a mean of 11.98 million euros. To better understand the variation of direct and indirect economic losses with PGA Figs. 5.11b, 5.12a and 5.12b, respectively shows the BI the DC and the total economic losses EL versus PGA. From Fig. 5.11b is evident that for PGA $\geq$ 1g the curves of 80th and 90th percentile reach a plateau of around 18–18.5 million euros, that becomes 16 million euros for the mean curve. Whereas for PGA ranging between 0.3 and 0.5 g the variation of BI is variable in mean between 0.4 and 0.7 millions euros. Figure 5.12a shows the

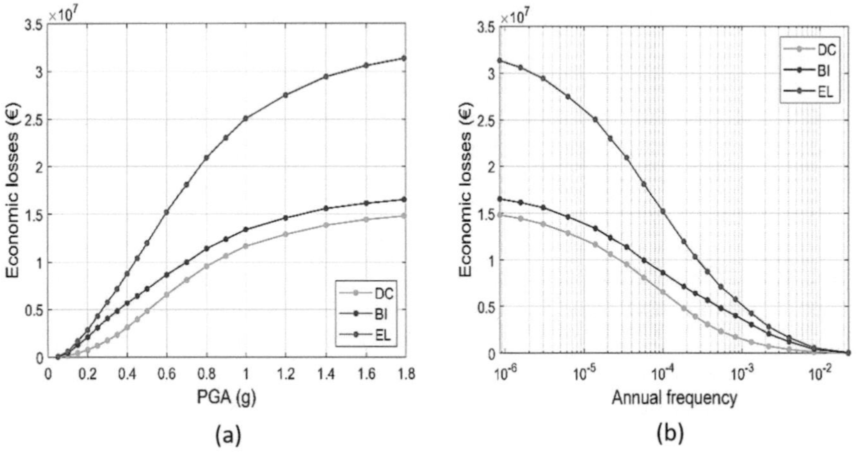

Fig. 5.13 **a** Direct cost losses for different PGA; **b** Probabilistic economic losses for different PGA

variation of DC for different levels of PGA. It can be noticed that the 50th percentile and mean curve have almost same values. The max values of DC between 10th and 90th percentiles, varies between 12.2 and 17.0 million euro. While the maximum DC losses for the mean curve are 14.8 million euro. Figure 5.12b plots the variation of EL with PGA. For PGA $> 1.2g$ the 80th and 90th percentile curves reach a plateau of around 32–35 million euros. Regarding the mean and 50th percentile curve, it can be notices that for PGA between 0.7 and 1 g the EL are almost same. On the contrary, for PGA smaller than 0.7g the mean EL are bigger than EL of 50th percentile curve. The maximum values of 10th and 90th percentiles values of EL is variable between 26 and 35 million euro. The maximum EL for the mean curve is about 31 million euro. In Fig. 5.13a the mean curves of DC, BI and EL of the black carbon plant for different levels of PGA are shown. For small values of PGA, (0.05–0.3 g) the BI is about 70–80% of EL, while for PGA levels >0.7 g the BI are slightly higher than DC. The mean expected loss exceedance curve is shown. In Fig. 5.13b. The mean expected annual losses (EALm) of the carbon black plant calculated with Eq. (5.6) is around 27.976 €, of which 19.599 € are due to business interruption and 8.378 € to the reconstruction of damaged equipment (direct costs). From this picture emerges that it is of paramount importance to design an efficient recovery plan in order to minimize the maximum recovery time and business interruption losses.

5.4 Conclusions

The consequences of catastrophic events on industrial districts can be extremely costly in terms of both fatalities and economic losses, also at community level. Thus, mitigation plans against natural hazards, the primary causes of such events, must be rethought. Indeed, the interaction between dangerous equipment and the presence of multiple damage conditions that characterise such events can amplify the effects of damage propagation. Moreover, the quantification of restoration capacity of the infrastructure and the community resilience represents a very difficult task. Natech events of dangerous industries, such as process plants, represents a particularly critical case that has been widely analyzed in the past, especially in presence of seismic loading. However, methods for quantifying the resilience of such complex systems are limited, and often do not account for all possible source of variability in quantifying hazard, vulnerability and reconstruction costs. For these reasons, this chapter proposed a full Probabilistic Resilience Analysis (PRA) for a quantitative evaluation the operational capacity and the direct and indirect economic losses of process plants generated by earthquakes. The method is an extension of the Performance-based Earthquake Engineering method, which also includes the restoration phase. The method, already proposed by the authors with a deterministic formulation, now incorporates the probabilistic nature of the problem. The proposed method takes into account the functional interactions between damaged equipment, for the quantification of the system's operational capacity over time, with a full probabilistic formulation. For this purpose, a simplified reconstruction network approach was proposed, by using empirical recovery costs and durations. The high number of random variables along with the multifaceted interaction between equipment and reconstruction activities, the method was formulated using a Monte Carlo approach. Consequently, by using a scenario-based approach, damage scenarios and reconstruction activities are sampled and statistical properties of the operational capacity and economic losses (direct and indirect) are derived. The proposed approach allows the plant managers to quantitatively derive resilience measures and individuate countermeasures for enhancing the performance of a process plant, optimizing also the related costs. A representative case study for the application and validation of the proposed method has been selected. In particular, a black carbon plant has been selected for the strategic role of such a facility and its complexity. The plant is ideally situated in Priolo Gargallo (Italy), which include several critical elements (storage tanks, pumps, stacks, piperack, etc.). According to the proposed method the following conclusions can be drawn:

- The operational capacity (OC) is quite scattered and dissymmetric with respect to the mean curve. The large dispersion of the recovery time of the equipment and the dominant damage states are the main caused of this behaviour. For instance, the 90th percentile of the OC curve, for PGA = 0.5 g, shows an out of scale overall recovery time (>720 d), evidently related to the delay caused by the recovery phase of damage scenarios including at least one vertical reactor, reactor and a vertical heat exchanger or the milling tower in damage state DS4, whose mean recovery time is 660 d.
- The dependency of the resilience index (R) from the intensity measure (PGA) is characterized by an initial zone where R is insensitive to PGA, that in mean, is extended up to PGA = 0.15 g with a 90% confidence interval between 0.1 and 0.3 g.

After that, R drops quite linearly up (in mean) to 1.3 g, becoming quite invariable for PGA > 0.8 g.

- The probability distribution functions of resilience index R and of recovery interval Tmax assumes a bimodal form, as shown for PGA = 0.5 g. This is essentially related to the presence, in the equipment, of different damage states and different recovery functions. In the example, the first peak (350 d) is characterized by prevalent damage states (DS4/DS5) in storage tanks and their recovery stages. The second peak is instead characterized by a prevalent DS4 damage state and recovery function in vertical reactors.
- The proposed approach allows a simple assessment of the variation of both direct (seismic damage to equipment, DC) and indirect (business interruption, BI) economic losses. For example, their probability distribution function, for PGA = 0.5 g, exhibited a clear prevalence of BI, higher, with respect to DC, of about 50%.
- The economic losses show a predominance of BI over DC up to 0.3g, after which the difference decreases and becomes <10% after PGA = 1.0 g. This means that for seismic intensities with return periods related to limit states typically considered for civil structures (up to collapse prevention), BI is the predominant cause of economic losses.
- The mean expected annual losses (EALm) indicate that more than 90% of the economic losses are related to seismic events with PGA <= 0.5 g, due to their higher frequency of occurrence. Furthermore, the BI represents 70% of the EALm. This indicates that it is of paramount importance to design an efficient recovery strategy to minimise economic losses.

Thus, the potential of the proposed method for decision making analysis of seismically vulnerable process plants is clear. The results in terms of operational capacity and economic losses can be profitably used to identify the most critical components of a plant and to investigate effective mitigation strategies, both in terms of vulnerability reduction and recovery time control. Further investigations will be focused on this aspect and the analysis of domino effects and dynamic interaction between equipment.

References

Abimbola, M., & Khan, F. (2019). Resilience modeling of engineering systems using dynamic object-oriented Bayesian network approach. *Computers & Industrial Engineering, 130*, 108–118. https://doi.org/10.1016/j.cie.2019.02.022

Alessandri, S., Caputo, A. C., Corritore, D., Giannini, R., Paolacci, F., & Phan, H. N. (2018). Probabilistic risk analysis of process plants under seismic loading based on Monte Carlo simulations. *Journal of Loss Prevention in the process Industries, 53*, 136–148. https://doi.org/10.1016/j.jlp.2017.12.013

Allin Cornell, C. (1968). Engineering seismic risk analysis. *Bulletin of the Seismological Society of America, 58*(5), 1583–1606. https://doi.org/10.1785/BSSA0580051583

Almufti, I., & Willford, M. (2013). Redi™ *rating system: Resilience-based earthquake design initiative for the next generation of buildings*. Arup Co. https://doi.org/10.13140/RG.2.2.20267.75043

Antonioni, G., Spadoni, G., & Cozzani, V. (2009c). Application of domino effect quantitative risk assessment to an extended industrial area. *Journal of Loss Prevention in the Process Industries, 22*(5), 614–624. https://doi.org/10.1016/j.jlp.2009.02.012

Ayyub, B. M. (2014). Systems resilience for multihazard environments: Definition, metrics, and valuation for decision making. *Risk Analysis, 34*(2), 340–355. https://doi.org/10.1111/risa.12093

Bakalis, K., Kohrangi, M., & Vamvatsikos, D. (2018). Seismic intensity measures for above-ground liquid storage tanks. *Earthquake Engineering & Structural Dynamics, 47*(9), 1844—1863. https://doi.org/10.1002/eqe.3043

Baker, J. W. (2008). *An introduction to probabilistic seismic hazard analysis (PSHA)* (Vol. 1, p. 72). White paper, version.

Bristow, D. N., Hay, A. H. (2017). Graph model for probabilistic resilience and recovery planning of multi-infrastructure systems. *Journal of Infrastructure Systems, 23*(3), 04016039. https://doi.org/10.1061/(ASCE)IS.1943-555X.0000338

Burnard, K., Bhamra, R., & Dani, S. (2011). Resilience: the concept, a literature review and future directions. *Int. J. Prod. Res., 49*(18), 5375–5393. https://doi.org/10.1080/00207543.2011.563826

Caputo, A. C., & Paolacci, F. (2017). A method to estimate process plant seismic resilience. In *Proceedings of the ASME 2017 Pressure Vessels and Piping Conference*, Volume 8: Seismic Engineering. V008T08A023. https://doi.org/10.1115/PVP2017-65464

Caputo, A. C., Kalemi, B., Paolacci, F., & Corritore, D. (2020). Computing resilience of process plants under Na-tech events: Methodology and application to seismic loading scenarios. *Reliability Engineering & System Safety, 195*, 106685. ISSN 0951-8320. https://doi.org/10.1016/j.ress.2019.106685

Chen, C., Reniers, G., & Khakzad, N. (2020). A thorough classification and discussion of approaches for modeling and managing domino effects in the process industries. *Safety Science, 125*, 104618. https://doi.org/10.1016/j.ssci.2020.104618

Cimellaro, G. P., Fumo, C., Reinhorn, A. M., & Bruneau, M. (2009). *Quantification of disaster resilience of health care facilities*. Technical Report MCEER-09-0009, University of Buffalo, NY, USA. https://www.eng.buffalo.edu/mceer-reports/09/09-0009.pdf

Corritore, D., Alessandri, S., Giannini, R., & Paolacci, F. (2017). Priamus: A new tool for the probabilistic risk assessment with monte carlo simulations of process plants under seismic loading. In *Proceedings of the ANIDIS Conference* (pp. 83–93). Pisa University Press. http://digital.casalini.it/4216203

Cruz, A. M., & Steinberg, L. J. (2005b). Industry preparedness for earthquakes and earthquake-triggered hazmat accidents in the 1999 Kocaeli earthquake. *Earthquake Spectra, 21*(2), 285–303.

Federal Emergency Management Agency. (2012). *FEMA p-58 seismic performance assessment of buildings*. Technical report, Federal Emergency Management Agency, Washington, DC. https://www.atcouncil.org/docman/fema/246-fema-p-58-1-seismic-performance-assessment-of-buildings-volume-1-methodology-second-edition/file

Gardoni, P., Sharma, N., & Nocera, F. (2021). Classification and mathematical modeling of infrastructure interdependencies. *Sustain. Resilient Infrastruct., 6*(1–2), 4–25. https://doi.org/10.1080/23789689.2020.1753401

Hazus FEMA. (2022). *Hazus earthquake model technical manual—Hazus 5.1*. Technical report, Federal Emergency Management Agency (FEMA). https://www.fema.gov/sites/default/files/2020-09/fema_hazus_earthquake-model_technical-manual_2.1.pdf

Hosseini, S., Al Khaled, A., & Sarder, M. D. (2016a). A general framework for assessing system resilience using Bayesian networks: A case study of sulfuric acid manufacturer. *Journal of Manufacturing Systems, 41*, 211–227. https://doi.org/10.1016/j.jmsy.2016.09.006

Hosseini, S., Barker, K., & Ramirez-Marquez,. J. E. (2016b). A review of definitions and measures of system resilience. *Reliability Engineering & System Safety, 145*, 47–61. https://doi.org/10.1016/j.ress.2015.08.006

Ivanov, D., Dolgui, A., Sokolov, B., Ivanova, M. (2017). Literature review on disruption recovery in the supply chain. *International Journal of Production Research, 55*(20), 6158–6174. https://doi.org/10.1080/00207543.2017.1330572

Kajitani, Y., Chang, S. E., Tatano, H. (2013). Economic impacts of the 2011 Tohoku-Oki earthquake and tsunami. *Earthquake Spectra, 29*(1_suppl), 457–478. https://doi.org/10.1193/1.4000108

Kalemi, B., Caputo, A. C., & Paolacci, F. (2019). Resilience calculation of process plants under seismic loading: A case study. In *Proceedings of the ASME 2019 Pressure Vessels and Piping Conference* (Vol. 5). V005T09A014. https://doi.org/10.1115/PVP2019-93311

Kalemi, B., Caputo, A. C., Corritore, D., & Paolacci, F. (2024). A probabilistic framework for the estimation of resilience of process plants under na-tech seismic events. *Bull. Earthq. Eng., 22*, 75–106. https://doi.org/10.1007/s10518-023-01685-z

Karaferis, N. D., Kazantzi, A. K., Melissianos, V. E., Bakalis, K., & Vamvatsikos, D. (2022). Seismic fragility assessment of high-rise stacks in oil refineries. *Bulletin of Earthquake Engineering*, 1–24. https://doi.org/10.1007/s10518-022-01472-2

Karagiannakis, G., Wenzel, M., Kowalczyk, P., Farhan, M., Zhelyazkov, A., Celano, F., Caprinozzi, S., Pedot, M., Kalemi, B., Bennani, H. K., et al. (2020). Seismic risk and resilience assessment of industrial facilities: Case study on a black carbon plant. In *17th World Conference on Earthquake Engineering, 17WCEE Sendai, Japan–September 13th to 18th 2020 Paper N° C003250*. https://www.researchgate.net/publication/345858334

Kazantzi, A. K., Bakalis, K., Vamvatsikos, D., Melissianos, V. E., Karaferis, N. (2022) An integrated model for the seismic risk assessment of an oil refinery. In *3rd International Conference on Natural Hazards & Infrastructure*, Athens, Greece.

Kazantzi, A. K., Karaferis, N. D., Melissianos, V. E., Bakalis, K., & Vamvatsikos, D. (2022). Seismic fragility assessment of building-type structures in oil refineries. *Bulletin of Earthquake Engineering*, 1–24. https://doi.org/10.1007/s10518-022-01476-y

Munoz, F., Mesa-Gomez, A., Casal, J. (2020). Risk analysis in Natech events: State of the art. *Journal Loss of Prevention in the Process Industries, 64*, 10471.

Mussini, N. (2019). *A multilevel decision-making approach for the resilience assessment of industrial plants in seismic prone areas*. Ph.D. thesis, Fakultät für Bauingenieurwesen der Rheinisch-Westfälischen Technischen Hochschule Aachen. https://doi.org/10.18154/RWTH-2018-226030

Nam Phan, H., Paolacci, F., Bursi, O. S., & Tondini, N. (2017). Seismic fragility analysis of elevated steel storage tanks supported by reinforced concrete columns. *Journal of Loss Prevention in the Process Industries, 47*, 57–65. https://doi.org/10.1016/j.jlp.2017.02.017

Nasiopoulos, G., Mantadakis, N., Argyroudis, S. A., & Mitoulis, S. A. (2021). Cost-based resilience assessment of bridges subjected to earthquakes. *International Journal of Disaster Resilience in the Built Environment, 12*(2), 209–222. https://doi.org/10.1108/IJDRBE-02-2020-0014

Nauck, F., Pancaldi, L., Poppensieker, T., & White, O. (2021). *The resilience imperative: Succeeding in uncertain times*. McKinsey Global Publishing.

Paolacci, F., Corritore, D., Caputo, A. C., Bursi, O. S., & Kalemi, B. (2018). A probabilistic approach for the assessment of LOC events in steel storage tanks under seismic loading. In *Proceedings of the ASME 2018 Pressure Vessels and Piping Conference*, Volume 8: Seismic Engineering. V008T08A016, https://doi.org/10.1115/PVP2018-84374

Paolacci, F., Giannini, R., & De Angelis, M. (2013). Seismic response mitigation of chemical plant components by passive control techniques. *Journal of Loss Prevention in the Process Industries, 26*(5), 924–935. ISSN 0950-4230. https://doi.org/10.1016/j.jlp.2013.03.003. Special Issue: Process Safety and Globalization.

Pasman, H. J., Keim, K. K., Mannan, M. S., Jain, P., & Rogers, W. J. (2018). A resilience-based integrated process systems hazard analysis (RIPSHA) approach: Plant system layer. *Process Saf. Environ. Prot., 116*, 92–105.

Pistikopoulos, E. N., Sammons, J., Eljack, F., Kazi, M. K., El-Halwagi, M. M., Sengupta, D. (2020). Disaster-resilient design of manufacturing facilities through process integration: Principal strategies, perspectives, and research challenges. *Frontiers in Sustainability, 1*, 595961. https://doi.org/10.3389/frsus.2020.595961.

Quiroz-Perez, E., Diaz-Ovalle, C., Mannan, M. S., Castillo-Borja, F., & Vazquez-Roman, R. (2018). A resilience index for process safety analysis. *J. Loss Prev. Process Ind., 50*, 184–189.

Reinhorn A. M., Bruneau, M., & Cimellaro, G. P. (2006) Quantification of seismic resilience. In *Proceedings of 8th U.S. National Conference on Earthquake Engineering*, San Francisco, California, USA. Paper No. 1094.

Sharma, N., G., Tabandeh, A., & Gardoni, P. (2019). Recovery optimization of interdependent infrastructure: A multi-scale approach. In *13th International Conference on Applications of Statistics and Probability in Civil Engineering, ICASP 2019*. https://doi.org/10.22725/ICASP13.148

Sheffi, Y. (2005). *The resilient enterprise: Overcoming vulnerability for competitive advantage*. MIT Press Books. https://mitpress.mit.edu/books/resilient-enterprise

Sheffi, Y. (2015). *The power of resilience: How the best companies manage the unexpected*. MIT Press. https://doi.org/10.7551/mitpress/9780262029797.001.0001

Sun, S., An, N., Wang, G., Li, M., & Zhou, J. (2019). Snap-back induced hysteresis in an elastic mechanical metamaterial under tension. *Applied Physics Letters, 115*.

Tong, Q., Yang, M., & Zinetullina, A. (2020). A dynamic Bayesian network-based approach to resilience assessment of engineered systems. *Journal of Loss Prevention in the Process Industries, 65*, 104152. https://doi.org/10.1016/j.jlp.2020.104152

Unnikrishnan, V., Gardoni, P., McAllister, T., van de Lindt, J., Guidotti, R., & Chmielewski, H. (2016). Modeling the resilience of critical infrastructure: The role of network dependencies. *Sustain. Resilient Infrastruct., 1*(3–4), 153–168. https://doi.org/10.1080/23789689.2016.1254999

Vanhoucke, M. (2012). *Project management with dynamic scheduling*. Springer. https://doi.org/10.1007/978-3-642-25175-7

Woods, D. D. (2015). Four concepts for resilience and the implications for the future of resilience engineering. *Reliability Engineering & System Safety, 141*, 5–9. https://doi.org/10.1016/j.ress.2015.03.018

Wu, Z., Lu, X., & Noori, M. (2020). *Resilience of Critical Infrastructure Systems: Emerging Developments and Future Challenges* (1st ed.). CRC Press. https://doi.org/10.1201/9780367477394

Chapter 6
Seismic Fragility Assessment of an Industrial Plant Substructure and Metastructures for Vibration Mitigation

Oreste S. Bursi

Abstract The tremendous impact of natural hazards, such as earthquakes, tsunamis, flooding, etc., which triggered technological accidents, referred to as natural-technological (NaTech) events, was demonstrated by several recent events as: (i) the Tohoku earthquake and the following Fukushima disaster in 2011; (ii) the Italian earthquakes in 2012 and 2016 which resulted in €42bn between direct and indirect costs, with casualties, thousands of displaced families, and affected business interruption that faced financial problems due non-existent or inadequate insurance. In addition to structural integrity, fitness for service of special risk industrial plants requires an understanding of specific threats, e.g. unrealistic design assumptions, insufficient controls at the construction stage, unexpected/unintended vibrations due to moving loads, etc. Therefore, measures to ensure safe operations depend on hazards and their relationships to the overall condition of the industrial plant. In this context, it is necessary to implement and support the Seveso-III Directive 2012/18/EU which regulates the control of major accident hazards involving dangerous substances and to extend the life and to ensure fitness for service of current special risk plants. To tackle this challenge the research conducted by the author and his coworkers involves a computational framework for an efficient and accurate seismic state-dependent fragility analysis. It is based on a combination of data acquired from an extensive experimental shake table test campaign on a full-scale prototype industrial steel frame structure and the most recent surrogate-based uncertainty quantification (UQ) forward analysis advancements. Specifically, the framework is applied to a real-world application consisting of seismic shake table tests of a representative industrial multistorey frame structure equipped with complex process components, carried out at the EUCENTRE facility in Italy, within the European SPIF project: Seismic Performance of Multi-component Systems in Special Risk Industrial Facilities. The results of this experimental research campaign also aspire to improve the understanding

O. S. Bursi (✉)
Department of Civil, Environmental and Mechanical Engineering, University of Trento, Via Mesiano 77, 38123 Trento, Italy
e-mail: oreste.bursi@unitn.it

V. Cozzani and F. Paolacci (eds.), *Management Strategies and Resilience Against NaTech Events*, CISM International Centre for Mechanical Sciences 616,
https://doi.org/10.1007/978-3-031-84485-0_6

of these complex systems and to improve the knowledge on FE modelling techniques. One goal aims to reduce the huge computational burden and to assess, as well, state-dependent fragility curves. Also innovative metamaterial-based solutions for disaster reduction in special risk facilities are proposed. To achieve this scope, the research builds upon the most recent innovative scientific and technological concepts developed within the research field of locally resonant phononic metamaterials. In particular, the feasibility of such metamaterial-based isolation concepts and systems is demonstrated across different low-frequency ranges, e.g. seismic waves. These general developments have further advanced through novel isolation concepts which can achieve negative effective mass and stiffness moduli in structures and through nonlinear mechanisms such as bistability. In sum, this research intends to conceive metastructures capable of rendering industrial plant equipment fully or partially isolated from the effects of incident waves impinging from all directions.

6.1 Introduction

6.1.1 Background and Motivation

Focusing on the industry world, from the last decade of major earthquake disasters, a key realisation has been that non-structural components (NSCs) of industrial plant facilities have been a significant driver for overall physical damage and disruption to business. To this scope, the Applied Technology Council (ATC) under the National Institute of Standards and Technology (NIST) invested in a year-long study to collect and document the body of available knowledge related to the seismic performance of NSCs, summarized in Rossi et al. (2020). From these studies, it clearly emerges: (i) a large adoption of classical load-and-resistance factor design (LRFD) and code-compliant methods often dated and inadequate, as highlighted also in Eads et al. (2013); (ii) a gap in performing fully comprehensive testing campaigns and investigations on coupling effects between primary structures and NSCs; (iii) a need to improve and enforce code requirements along with the development of reliable NSCs seismic demands models. In particular, codes and regulations establish the minimum requirements for structures to resist the effects of earthquakes with defined probability levels (or return periods), providing a basic level of seismic risk reduction. The primary objective of codes and regulations on structures is to save lives, not the structures themselves. When properly implemented, codes have significantly reduced the loss of life in major seismic events in recent years. However, from the perspective of economic resilience, the enormous scale of associated financial losses requires a paradigm shift. Encouragingly, in recent times, earthquake building codes have been revised to implement elements of a performance-based earthquake engineering (PBEE) framework to address issues of damage limitation and reparability as well as life safety. In this respect, the PBEE PEER framework has gained momentum above all to perform seismic risk analyses, thanks to its inherent versatility. Its strength, indeed, relies on simple yet effective implementation of the total probability theorem, which allows to decouple and then combine the output of probabilistic seismic hazard analysis (PSHA) with fragility, damage, and loss analysis. To this end, the aforementioned fragility analysis step offers the critical link between seismic hazard and structural modelling, since it estimates the conditional probability of exceeding or attaining a certain damage state (DS) given an intensity measure (IM) of earthquake motion. In this regard, a growing number of studies, the increase of community-funded research projects and experimental campaigns focused on unfolding the complex and highly diverse behaviour of these process components. For example, the European research project INDUSE-2-SAFETY 2012-2015, investigated through experimental tests the performances of several NSCs and their limit state levels to be compliant with. Conversely, the United States Nuclear Regulatory Commission agency carried out a multi-year test programme to develop a better understanding of the elastoplastic response and ultimate strength of nuclear power piping systems. As a result, in view of the enhancement of structural integrity of industrial, energy and nuclear components as well as fitness for service, design guidelines and recommendations were issued (NISTGCR 2017).

Despite the available technical solutions, earthquakes remain one of the most harmful natural hazards in industrial plants due to domino effects. The applicability of conventional base isolation is limited to specific exceptional cases mainly because seismic isolation devices exhibit a limited life cycle time due to certain components -e.g. rubber bearings, friction surfaces, seals, etc.-. Recently, innovative scientific and technological concepts have been developed within the broad field of metamaterials (Maldovan 2013). These are natural or artificial materials or structures which exhibit extraordinary properties for inhibiting or conditioning wave propagation in all spatial directions over broad frequency bands. Such materials are already finding successful technological applications for high-frequency waves, i.e. electromagnetic, optical, thermal and ultrasound. Quite recently, the application of metamaterial concepts has also been proposed for seismic and sound waves (Palermo et al. 2016; Basone et al. 2019a; Miniaci et al. 2016; Hussein et al. 2014). These initial developments have further advanced through novel absorption and isolation concepts, such as the addition of negative effective mass and stiffness devices, and periodic structures endowed with nonlinear mechanisms, e.g. bistability (Frazier and Kochmann 2017; Hussein and Frazier 2013; Bursi et al. 2021a). In this way, the limitations of existing seismic and low-frequency isolation technologies can potentially be overcome, leading to highly-damped, low-maintenance and easily implemented designs based on locally resonant metabarriers and metastructures capable of shielding critical equipment of industrial facilities. Similarly, new devices like inerters and nonlinear impact-based energy sink devices (NES) have been recently used in metastructures for mass reduction and free vibration mitigation (Rothos and Vakakis 2009; Wenzel et al. 2020).

Along these lines, the author has tried to integrate and exploit the new capabilities offered by metamaterials and metastructures, i.e. metafoundations, to expand standard safety systems e.g. seismic isolation; see, in this respect, Fig. 6.1.

With reference to broad tanks subjected to seismic loading, instead, relevant fragility functions were determined by means of 3D FE models and low-fidelity

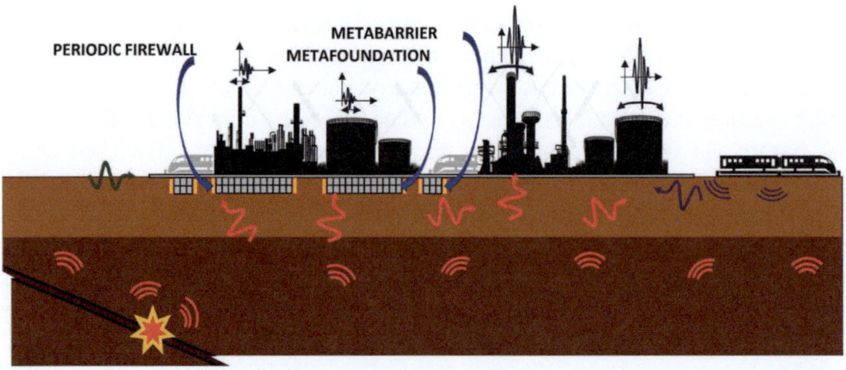

Fig. 6.1 Reduction of vibrations due to seismic loadings and other hazards through the use of metastructures in a petrochemical facility

kriging models based on Gaussian process regression (Phan et al. 2020) the stochastic nature of the input was taken into account. Finally, both aleatoric and epistemic uncertainties with their propagation in the design/modelling of metamaterials and metastructures still need to be examined in depth (Celli et al. 2019).

One of the important challenging points of utilizing MF is the excessive size necessary to obtain an operational design and the material strength and technology limitations to reach optimum stiffness and damping requirements. For the solution, many different approaches and techniques including nonlinearities and novel devices have been proposed. In the case of heavy or low-frequency superstructure, the resonators need to be tuned at ultra-low frequency ranges, which can be solved with the use of negative stiffness elements in parallel with the wire ropes that attach the resonators to the main cell (Wenzel et al. 2020). Bursi et al. (2021b) investigated the wire ropes in the nonlinear range using hysteretic Bouc-Wen model and proposed an optimization method. Achieving proper attenuation necessitates the installation of substantially heavy resonators. One alternative approach to achieve mass amplification is by incorporating inerter devices. Inerters respond with a force that is directly proportional to the relative acceleration between their two terminals (Smith 2002). The significant advantage of using inerters lies in their ability to possess inertance levels several orders of magnitude higher than the actual mass. As a result, the inertance term -b- does not factor into the equation of motion for the dynamic system under the influence of ground motion. In the field of structural engineering, many researchers have proposed the use of inerters to enhance seismic protection (Ma et al. 2021; Liu et al. 2022). Giaralis and colleagues conducted studies on Tuned Mass Dampers (TMDs) equipped with inerters, exploring their effectiveness in suppressing vibrations caused by both wind and earthquake excitations (Marian and Giaralis 2017; Giaralis and Petrini 2017).

6.1.2 Scope

The first part of this work presents an innovative, efficient and accurate framework for the evaluation of state-dependent fragility functions for critical industrial process components. The relevant vulnerability assessment is based on the integration of data derived from an intensive experimental campaign, accomplished within the European SERA SPIF project, with the computational potential obtained from advanced meta-modelling techniques. The value of this framework resides in the experimental data on which it is built. These are the outcome of the comprehensive experimental programme completed at the EUCENTRE laboratories in Pavia (I). More precisely, the tests were carried out on a shaking table of two archetypes of industrial systems full-scale steel substructures, equipped with NSCs such as tanks, cabinets, BFJs, and piping elements. In fact, experimental data fill the gap in terms of experimental evidence of tests conducted in a PBEE framework capable of taking into account the effects of NSCs-primary structure coupling as a whole. Therefore, an added value of this work lies in the performance evaluation and study of the complex interactions between NSCs and the primary structures through the development of different highly refined numerical models capable of capturing interaction effects. They are conceived to be more quick and inexpensive-to-run and able to capture the global behaviour of each coupled system. For the sake of brevity, we'll concentrate on the braced frame.

Moreover, this chapter presents a novel seismic protection method for industrial process components. This method is based on the core concept of locally resonant metastructures, specifically termed finite locally resonant metafoundations (MFs). The specific objectives pursued to achieve this goal include: (i) the identification of parameters of MFs and possible benefits of introducing nonlinear dynamics to MFs to enhance the performance; (ii) the investigation of more efficient strategies to model superstructures for optimization and design of MFs; (iii) to ensure code-compliant design for MFs and to develop efficient optimization strategies to determine optimal resonator parameters; (iv) the implementation of bistable columns in metafoundations to leverage advantageous features of bistability such as harmonic energy diffusion and adaptive bandgaps, and to assess their impact on the characteristics of finite lattices and periodic systems. In addition, we underline the scientific and technical feasibility of the proposed solutions. Finally, conclusions and future perspectives are drawn.

6.2 Dynamic Performance of a Multistorey BF Frame Structure with Process Equipment Subjected to Site-Based Ground Motion

The problem complexity and the inherent modelling of plants concur in limiting the development of risk assessment methods, such as component and system fragility, based on meaningful reduced and quick-to-run models. FE-based and surrogate models, such as Kriging, can then be used to estimate both component and system fragility functions of process equipment coupled to the primary structures.

6.2.1 A New Framework for System Vulnerability Assessment

Due to constraints in time and available information, risk-informed assessments are often conducted using simplified, quick-to-evaluate models. However, as emphasized in the recent NISTGCR 17-917-44 (NISTGCR 2017) report on industrial facilities, these established approaches do not result in economically viable or rational designs. Building on this, the current subection introduces the concept of a new framework for system vulnerability assessment. This framework leverages a combination of experimental data and surrogate models, facilitating the computation of state-dependent fragility curves that account for multiple aspects of the problem. State-dependent fragility curves are a specific type of fragility curve that are conditioned not only on a measure of seismic intensity (IM), which is typically represented as a vector, but also on the initial state of discrete damage (DSi) of the structure (see Iervolino et al. 2015b). These curves therefore allow for the evaluation of a system's vulnerability after it has already sustained damage, as defined by the generalized equation,

$$\begin{aligned}\mathbb{P}\left[\mathrm{DS_j}|\mathrm{DS_i},\ \boldsymbol{IM}=\boldsymbol{im}\right]&=\mathbb{P}\left[\mathrm{DS}\geq\mathrm{DS_j}|\mathrm{DS_i},\ \boldsymbol{IM}=\boldsymbol{im}\right]+\\&-\mathbb{P}\left[\mathrm{DS}\geq\mathrm{DS_{j+1}}|\mathrm{DS_i},\ \boldsymbol{IM}=\boldsymbol{im}\right],\end{aligned}\tag{6.1}$$

for $j > i$, with i indices representing the identified damage limit states. Figure 6.2a illustrates the generalized transition probability state matrix for a system with three possible levels of damage -labeled DS0 to DS2-; whilst Fig. 6.2b depicts the associated Markovian diagram transition states.

As a consequence, the derivation of state-dependent fragility functions inherently lies in a vast number of Monte Carlo analyses. However, the complexity of the FE models hinders the direct derivation. Hence, an innovative UQ framework based on a reduced number of expensive FE simulations and on leveraging cutting-edge surrogate model techniques is proposed. Specifically, Fig. 6.3 outlines the UQ-based framework and the key steps for performing state-dependent fragility analysis. The top row, highlighted with a red contour, represents the brute-force MCS

Initial damage state DS_i \ Transition to damage state DS_i	$\mathbf{DS_0}$	$\mathbf{DS_1}$	$\mathbf{DS_2}$
$\mathbf{DS_0}$	$1 - \mathbb{P}_{01} - \mathbb{P}_{02}$	$\mathbb{P}[DS \geq DS_1 \mid DS_0, IM] - \mathbb{P}[DS \geq DS_2 \mid DS_0, IM]$	$\mathbb{P}[DS \geq DS_2 \mid DS_0, IM]$
$\mathbf{DS_1}$	×	$1 - \mathbb{P}_{12}$	$\mathbb{P}[DS \geq DS_2 \mid DS_1, IM]$
$\mathbf{DS_2}$	×	×	1

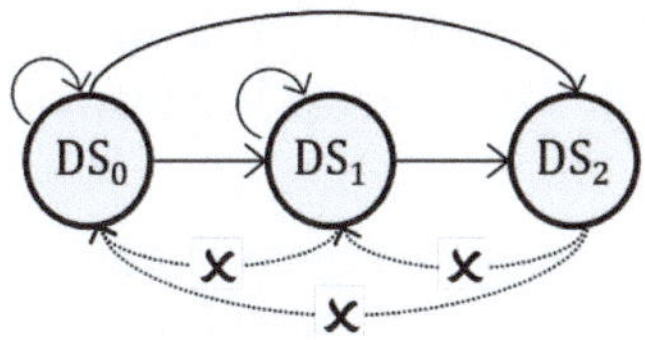

Fig. 6.2 Transition state: **a** matrix and **b** diagram, respectively

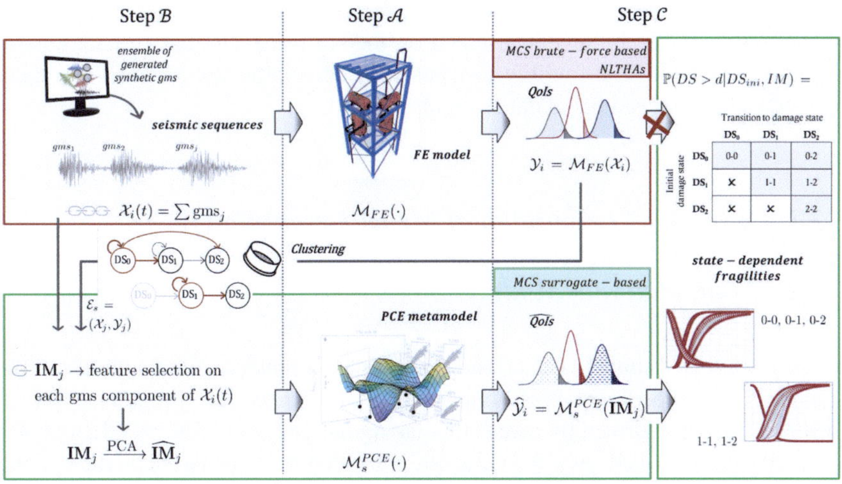

Fig. 6.3 UQ-based framework and key steps for performing a state-dependent fragility analysis

approach using computationally expensive NLTHAs (steps $\mathcal{A}$ to $\mathcal{C}$), whilst the green-highlighted row illustrates the surrogate-based MCS method. In the top row, step $\mathcal{A}$ involves developing complex and computationally intensive FE models, followed by step $\mathcal{B}$, where a ground motion model is used to generate stochastic seismic sequences as input. However, the heavy computational load limits the feasible number of MCS runs, creating a constraint on the derivation of state-dependent fragility functions, as indicated by the interrupted red-crossed arrow. Despite this, the limited NLTHA QoIs obtained are clustered based on the initial damage state metric DS0,1,2. These clustered data, along with the corresponding seismic event sequences, form the DoE Ds for the surrogate models (steps $\mathcal{A}$ and $\mathcal{B}$). First, a comprehensive vector of IMs is extracted, and then a principal component analysis (PCA) is applied to reduce the input dimensionality. This reduction allows for the construction of polynomial chaos expansion (PCE) metamodels for each Ds, where s ∈ [DS0, DS1]. Finally, a large number of surrogate-based MCS runs are conducted, enabling the derivation of state-dependent fragilities (step $\mathcal{C}$).

6.3 Industrial Case: A Vertical Tank Installed on a Special-Risk 3D Braced Structure

The assessment of system fragility functions is herein presented for the industrial mock-up of the project SPIF #2, i.e., *Seismic Performance of Industrial Facilities*. Specifically, the project consisted of a comprehensive experimental shake table test campaign on an archetypal multi-component steel braced frame, endowed with industrial process components. A state-dependent fragility analysis was performed on one of the most critical equipment installed on the structure, i.e., the vertical tank at the first-floor level. The fragility analysis focused on this component, leveraging bootstrap-PCE and campaign data insights.

6.3.1 Step $\mathcal{A}$—Computational Model Description

6.3.2 Computational Models

FE physics-based model. The considered FE model represents the outcome of the recent SPIF #2 experimental research campaign, i.e. *Seismic Performance of Industrial Facilities*, performed by means of shaking table tests at EUCENTRE Laboratory, Italy. Figure 6.4a depicts the industrial substructure model tested on the shaking table. The mock-up consists of a full-scale 3-storey steel braced frame (BF) structure with flexible floors. Several NSCs of the industrial process were installed, like tanks, cabinets, bolted flange joints, and T-joints. Particularly, the complex dynamic interaction between the main steel structure and one of the NSCs—the vertical tanks mounted on the first level—was of great interest. The strong rocking exhibited by those components deserved attention due to the possible disastrous consequences related to damage on them. Figure 6.4a shows the investigated vertical tank. Careful notice was paid to the FE modelling. In this respect, Fig. 6.4b reports a detail of the ad-hoc designed stick model for the tanks and its expensive reference high-fidelity local FEM. The stick model was developed with the goal to: (i) mimic the modal properties and: (ii) catch the different effects of the participating seismic mass on the supporting girders system, varying with the intensity level of the seismic excitation. A thorough discussion on FE modelling and calibration of the global SPIF #2 model of Fig. 6.4c is deferred in Quinci et al. (2023). This was accomplished through the

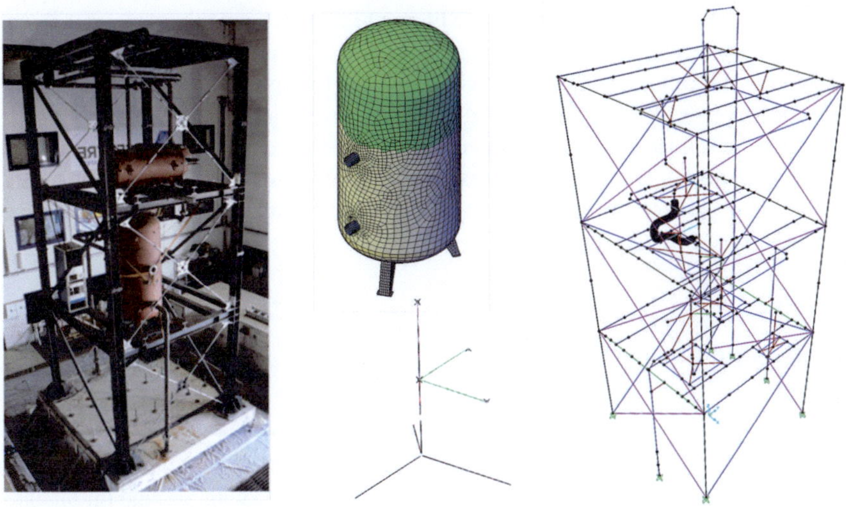

(a) (b) (c)

Fig. 6.4 The SPIF #2 mock-up: **a** photo of the braced frame (BF) configuration on the shake table of EUCENTRE Facilities and details of the vertical tank installed at the first level; **b** SAP2000®high-fidelity local FEM and ad-hoc implemented stick-model for the vertical tank; **c** global SAP2000®FEM of SPIF #2

deployment of complex and time-consuming high-fidelity local FE models, which were then used to assemble and calibrate the global physics-informed FE model in Fig. 6.4b.

The shake table tests showed that among the installed components, the vertical tanks were found to be critical and of greater interest. Therefore, fragility analysis is focused on these elements, leveraging bootstrap-PCE and campaign data insights.

6.3.3 Step $\mathcal{B}$—Input Definition

To simulate different damage initial configurations and cover each transition state, sequences of seismic events were assigned as input to the NLTHAs of the FEM of SPIF #2. A total of 100 gms sequences, consisting of chains of 5 earthquakes each, were cast by the same ensemble of gms generated by the stochastic ground motion model of Rezaeian and Der Kiureghian (2010). The computational cost pro single sequential NLTHA is ~30 min on an Intel(R) Core(TM) i9-10900K CPU @3.70GHz, 10 Core(s)—128 GB RAM. Both the number of gms in a single sequence and the total number of sequential NLTHAs were determined by a trade-off between the total time required for analysis and a well-defined population of the transition matrix of Fig. 6.2b. The transition matrix for the vertical tank was defined on two performance limit states. Those were inferred by the literature review, see Vathi et al. (2017), and confirmed by the experimental campaign, see Nardin et al. (2022). The first was the Design Basis Earthquake (DBE), which is linked to the operation and functionality of the process plant. The second was the Safe Shutdown Earthquake (SSE) limit state, for which the fundamental safety functions can be ensured, although the facility is no longer operational. Thresholds were experimentally identified in Nardin et al. (2022) as maximum acceleration recorded at the base of the tank: 6.72 m/s^2 for DBE and 18.5 m/s^2 for SSE, respectively.

6.3.4 Step C—QoI Response

NLTHAs and experimental design generation. The results of the NLTHAs performed on the FE model were clustered into the identified six transition states of Fig. 6.2a. Specifically, DS_0 indicates the dataset associated with pristine initial damage conditions, whilst DS_1 denotes the dataset for which the threshold associated with the DBE limit state was exceeded. Lastly, DS_2 collects the results of simulations for which the SSE limit state was attained. Almost 47% of the simulations belong to the DS_0 damage state initial condition. Particularly, 13 and 7% transitioned from undamaged initial conditions to exceeding the DBE and SSE threshold at the end of the NLTHAs, respectively. Instead, 38% of simulations belong to DS_1, out of which 16% stayed in the same damage level even after other seismic shocks. Finally, 15% of the simulations reached the absorption state. Finally, in agreement with Table 6.1, 41 IMs were evaluated for each gms of the NLTHA simulations. To reduce the order of the input dimension, PCA was then applied to each gms. To achieve 99% of the variability of data, 10 PCs were needed.

PCE metamodelling. State-dependent fragilities required many more analyses than the ones performed on the expensive-to-run FE model. Hence, two metamodels were implemented to overcome the computational and time issues. Specifically, the first metamodel was built on the experimental design $\mathcal{E}_0$ of simulations belonging to the DS_0 dataset, i.e. with undamaged initial conditions. The second had as experimental design $\mathcal{E}_1$, the family of simulations belonging to DS_1, i.e., with damage initial conditions attaining the DBE limit state. Resampling with substitution was used on both the experimental design $\mathcal{E}_{0,1}$, thus generating sets of bootstrap replications. Due to the finite size of the experimental design, such sets of responses were deployed to calculate the statistics of interest, i.e., the quantiles of the PCE predictor. Hence, bootstrap PCE (bPCE) was carried out with a q-norm truncation with $q = 0.50$, maximum allowed interaction $r = 2$ and a number of replications $B = 500$. Based on the performance of the ε_{LOO} and εemp errors, the subspace pursuit (SP) solver performed best, as in the benchmark test. As a result, Fig. 6.5a reports the histogram of the surrogate predictor over the histogram distribution of the $\mathcal{E}_0$. The limit state thresholds of DBE and SSE are also reported. A favourable performance is attained, with a final ε_{LOO} error estimate of 4.17E-02. The estimation of bPCE coefficients

Table 6.1 A comprehensive list of ground motion IM parameters (De Biasio et al. 2015; Hariri-Ardebili and Saouma 2016

No.	Description of IM	Symbol	Mathematical model
1	Peak ground acceleration	PGA	$\max(\lvert\ddot{u}(t)\rvert)$
2	Peak ground velocity	PGV	$\max(\lvert\dot{u}(t)\rvert)$
3	Peak ground displacement	PGD	$\max(\lvert u(t)\rvert)$
4,..,8	Spectral displacement	Sd	$S_d(T_x)$
9,..,13	Spectral velocity	Sv	$S_v(T_x)$
14,..,18	Spectral acceleration	Sa	$S_a(T_x)$
* *where* $T^* \in [0.50; 0.35; 0.25; 0.15; 0.10]$			
19	Arias intesity	$\mathrm{I_A}$	$\pi/2g \cdot \int_0^{t_f} a^2(t)\mathrm{d}T$
20	Total cumulative energy	$\mathrm{E_{cum}}$	$\int \ddot{u}^2(t)\mathrm{d}t$
21	Riddell–Garcia intensity acceleration	$\mathrm{I_{RG,a}}$	$(PGA)^{+1} \cdot (T_d)^{+1/3}$
22	Riddell–Garcia intensity velocity	$\mathrm{I_{RG,v}}$	$(PGV)^{+2/3} \cdot (T_d)^{+1/3}$
23	Significant time duration	$\mathrm{T_d}$	$t_{95} - t_{05}$
24	Root mean square of acceleration	RMS($\ddot{u}$ (t))	$\sqrt{(\frac{1}{N}\sum_{n=1}^{N}\lvert x_N^2\rvert)}$
25	Characteristic intensity	IC	$\mathrm{RMS}(\ddot{u}(t))^{1.5} \cdot T_d^{0.5}$
26	Cumulative absolute velocity	CAV	$\int_0^{t_f} \lvert a(t)\rvert\mathrm{d}T$
27	Cosenza–Manfredi intensity	$\mathrm{I_{CM}}$	$2g/\pi \cdot (PGA)^{-1}(PGV)^{-1}(AI)^{+1}$
28	Average spectral acceleration	$\mathrm{ASA_{40}}$	$2.5/f_1 \int_{0.6 \cdot f_1}^{f_1} S_a(f,\varepsilon)\mathrm{d}f$
29	Acceleration spectral intensity	ASI	$\int_{0.1}^{0.5} S_a(T,\varepsilon)\mathrm{d}T$
30	Effective peak acceleration	EPA	$1/2.5 \int_{0.1}^{2.5} S_a(T,\varepsilon)\mathrm{d}T$
31	Velocity to acceleration ratio	$\mathrm{I_{v/a}}$	PGV/PGA
32	Fajfar intensity	$\mathrm{I_F}$	$(PGV)^{+1} \cdot (T_d)^{+1/4}$
33	Mean frequency	$\mathrm{F_m}$	$\sum_i U_i^2(f_i)/\sum_i U_i^2$
34	Rate of change mean frequency	$\dot{F}_m$	$\mathrm{d}F_m(T)/\mathrm{d}t$
35	Fourier amplitude spectrum area	$\mathrm{FAS_{area}}$	$\frac{1}{4df}\int_{f_1-2df}^{f_1+2df} U(f)\mathrm{d}f$
36,..,41	Equipment relative average spectral acceleration	$\mathrm{E-ASA^{**}_{R_xX}}$	$\frac{1}{f_1 \cdot (X_f-1)} \cdot \int_{f_1}^{X_f \cdot f_1} S_a(f,\varepsilon)\mathrm{d}f$

** R indicates the chosen percentage of drop of the fundamental frequency ($X_f = 1 - (R/100)$); $R \in [40; 67; 80; 100; 150; 200]$

converged to a polynomial degree order of 3. Moreover, Fig. 6.5b shows the control $Y_{\mathcal{E}_0} - Y_{bPCE}$ plot: a good alignment is found with the 45° line, which represents the ideally perfect match between true and surrogate data.

Similarly, bPCE was performed for the experimental design $\mathcal{E}_1$. Anew, a q-norm truncation with $q = 0.50$, maximum allowed the interaction $r = 2$ and a number of replications $B = 500$ was applied. According to the εemp error, the SP solver was identified as the best one. However, the optimal matching pursuit OMP solver also performed significantly well regarding the εLOO error. As a result, Fig. 6.6a reports

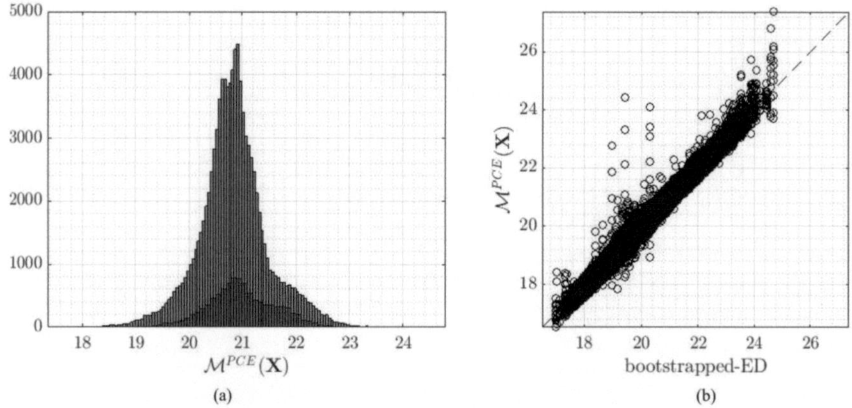

Fig. 6.5 **a** Histogram distribution of the PCE surrogate Y_{PCE} (light gray) predictor *vs* the $Y_{\mathcal{E}_0}$ original data of the initial undamaged condition dataset; **b** control plot of the performance of the Y_{bPCE} surrogate model *vs* the $Y_{\mathcal{E}_0}$ reference experimental design samples

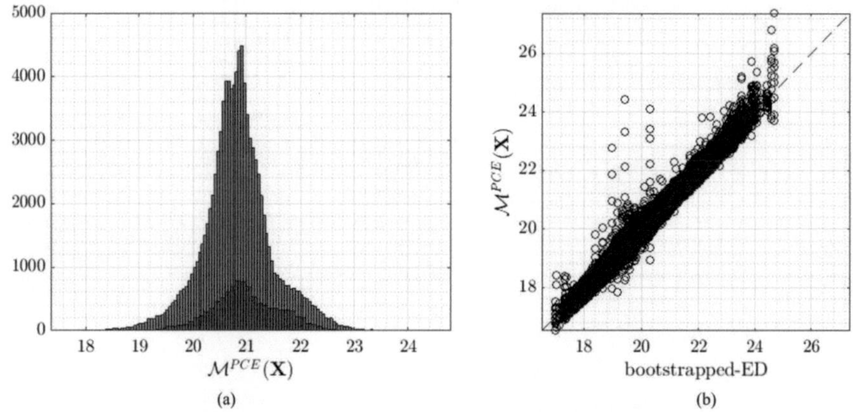

Fig. 6.6 **a** Histogram distribution of the PCE surrogate Y_{PCE} (light gray) predictor *vs* the $Y_{\mathcal{E}_1}$ original data of the initial undamaged condition dataset; **b** control plot of the performance of the Y_{bPCE} surrogate model *vs* the $Y_{\mathcal{E}_1}$ reference experimental design samples

the histogram distribution of both the Y_{PCE} predictor and the $Y_{\mathcal{E}_1}$ data of the damaged initial state condition. A generally good agreement is found, with a few exceptions in the range between [21 and 22] m/s^2. This is reflected also in the control plot Y_{bPCE}–$Y_{\mathcal{E}_1}$ of Fig. 6.6b, where the dispersion is greater in that range. The estimation of PCE coefficients converged at a polynomial degree of order 5 with a final ε_{LOO} error of 4.73E-02.

Fragility assessment. To determine which IM is better, the efficiency index β_{eff} is evaluated for each transition state. Particularly, the optimal IM correspond to the one whereby the variability of the EDP is minimal. Consequently, for each IM, the efficiency index β_{eff} was evaluated as follows:

$$\beta_{\text{eff}}(im) = \min\mathbb{E}\left(\text{IQR}|_{im,x_1} + \text{IQR}|_{im,x_2} + \text{IQR}|_{im,x_3}\right) \tag{6.2}$$

where IQR stands for the 90th inter-quantile range evaluated in correspondence of the 25th–50th–75th quantile of the IM axes. The lower the β_{eff} index, the better the performance of the IM. Specifically, the PGA, E_{cum}, and $\text{E} - \text{ASA}_{\text{R67}}$ are the optimal IM that occurred more often. Those are positively correlated with acceleration and energy content; see, in this respect, the definition reported in Table 6.1; whilst, another IM that often recurred, was the I_{F}, correlated with velocity terms. As thoroughly investigated in the research paper (Fajfar et al. 1990), the I_{F} captures damage for structures with fundamental periods in the medium-period (velocity-controlled) region. Instead, the $\text{E} - \text{ASA}_{\text{R}_{67}}$ IM captures the presence of significant spectral acceleration ordinates over the structure's dominant frequency interval, as reported in De Biasio et al. (2015). Therefore, it accounts for damage by detecting frequency shifts due to structure-component coupling effects. These observations agree with the experimental evidence described in Nardin et al. (2022). More precisely, the processing of the collected data revealed a significant correlation between the maximum floor spectral acceleration $Sa_{floor}(T_{NSC})$, evaluated at the identified period $T \simeq 0.16$s of the vertical tanks and the $\text{E} - \text{ASA}_{\text{R}}$ measure.

Figure 6.7 reports the state-dependent fragility curves as functions of the PGA. In the Fig. 6.7, dark-red lines and the associated shaded areas highlight the 50, 90 and 99% confidence bounds, respectively. Given an initial damage condition, i.e., -graphically- for elements of the same row, the variability increases towards higher damage states, i.e., moving to the right. Similarly, given a damage limit state to attain, i.e., -graphically- for elements of the same column, the required magnitude of IM to reach the same exceedance probability is lower. For instance, the 50% probability of exceedance DS_2 given DS_1 and DS_0 is associated to PGA = 2 and PGA = 3.7 m/s^2, respectively.

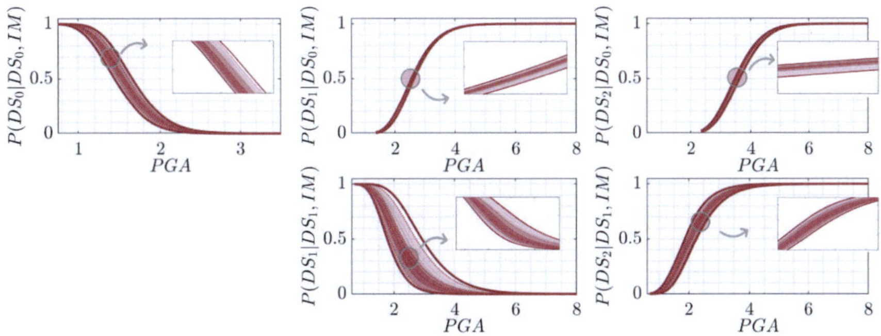

Fig. 6.7 Bootstrapped-PCE state-dependent fragility curves of the SPIF #2 vertical tank with PGA as IM: dark-red thick lines stand for the 50, 90 and 99% confidence bound, along with lighter to darker shaded areas

6.4 Numerical and Experimental Activities on Bistable Components and Metamaterials

6.4.1 Numerical Characterization of Bistable Systems

Bistable oscillations have garnered significant attention in the last decade due to their sophisticated and diverse dynamics. Bistability is characterized by a system with two stable equilibrium states, and is observed in biological systems like genetics (Dubnau and Losick 2006) and neuroscience (Knapen et al. 2011; Durstewitz and Seamans 2006). In the mechanical field, buckled columns or beams serve as examples of bistable systems (Cazottes et al. 2009). Therefore, the advantages of bistable oscillations have entailed wide applications in various phenomena, including energy harvesters (Harne and Wang 2013; Arrieta et al. 2010), nonlinear energy sinks (Habib and Romeo 2017), and MEMS (Saif 2000).

Bistable systems exhibit a complex feature known as stochastic resonance (Gammaitoni et al. 1989), characterized by snap-through dynamics induced by low-level noise and periodic excitation. Interestingly, individually, both low-level noise and periodic excitation are insufficient to trigger interwell dynamics (snap-through motions between two stable configuration), making bistable systems beneficial for energy harvesting. Furthermore, the presence of strong nonlinearity in these systems leads to chaotic responses, as per chaos theory, revealing underlying patterns and deterministic laws within the randomness of chaotic dynamics (Hopf et al. 1982). The butterfly effect (Lorenz 2000), characterized by extreme sensitivity to initial conditions, creates diverse and intricate dynamic response characteristics in systems enodwed with bistability.

The basic static nonlinearity exhbited by a bistable component can be obtained through a Duffing oscillator with a negative first-order and a positive third-order spring constants. In particular, the static force-deformation and potential energy relationships of a bistable Duffing oscillator with ($P_1 < 0$, $P_3 > 0$), is given in Fig. 6.8, where the positive and negative stiffness regions can be clearly identified. In the energy graph, the two stable points, the so called stable wells can be observed, and the distance between the cells corresponds to $\sqrt{|P_1/P_3|}$.

At the stable points, the Duffing spring is characterized by a stiffness value,

$$\left(P_1 x + P_3 x^3\right) \frac{d}{dx}\bigg|_{x=\pm\sqrt{-P_1/P_3}} = -2P_1 \tag{6.3}$$

whilst between the stable wells, the system exhibits softening, until reaching the negative stiffness region where the maximum absolute value of the negative stiffness will be reached at the unstable point, i.e. $x = 0$; at this point, the stiffness value reads P_1. When the dynamics are considered, in the most general terms the motion of the system can be divided into two parts: the intrawell motion and the interwell motion. In Fig. 6.9a in particular, the intrawell motion is given together with the phase portrait where the energy of the system is not sufficient to breach over the unstable point

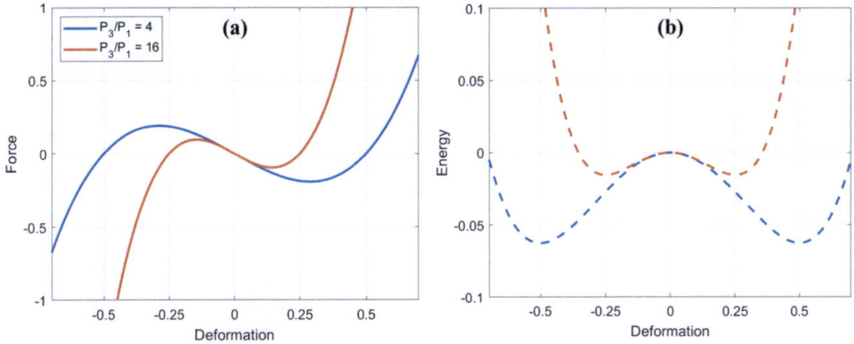

Fig. 6.8 Bistable Duffing oscillators: **a** force-deformation relationships; and **b** Relevant potential energies

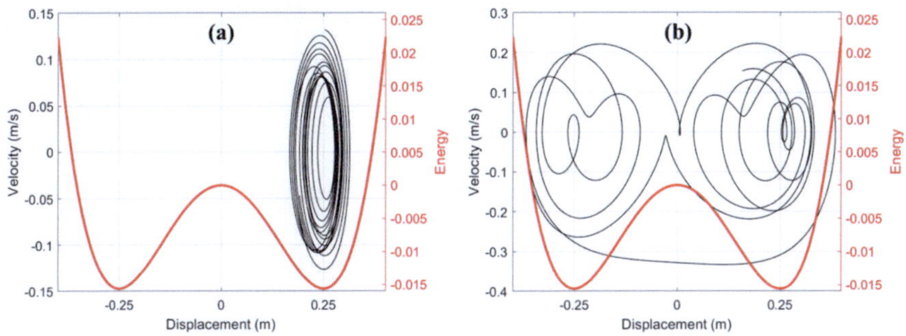

Fig. 6.9 Bistable Duffing oscillator motions in phase portrait: **a** Intrawell motion; and **b** Interwell motion

energy and the system oscillates around one of the stable wells. In Fig. 6.9b instead, the system has sufficient energy to overcome the potential peak in the middle; then, interwell motions can be observed, where oscillations happen between and around wells. These motions can also be distinguished according to their amplitude in small and large amplitude.

A bistable lattice composed of bistable undamped springs and masses is depicted in Fig. 6.10, and the EoM of the system can be written as,

$$m_n\ddot{x}_n + P_1(2x_n - x_{n-1} - x_{n+1}) + P_3(x_n - x_{n-1})^3 + \dots$$
$$\dots + P_3(x_n - x_{n+1})^3 = 0 \quad (6.4)$$

where $P_1 < 0$, $P_3 > 0$. Herein, it must be noted that the harmonic balance method (HBM) is not suitable for systems with aperiodic responses; as a result, an analytical solution was not pursued. The plot of the numerically obtained dispersion results are presented in Fig. 6.10 with various amplitudes. In the vicinity of dissipation and

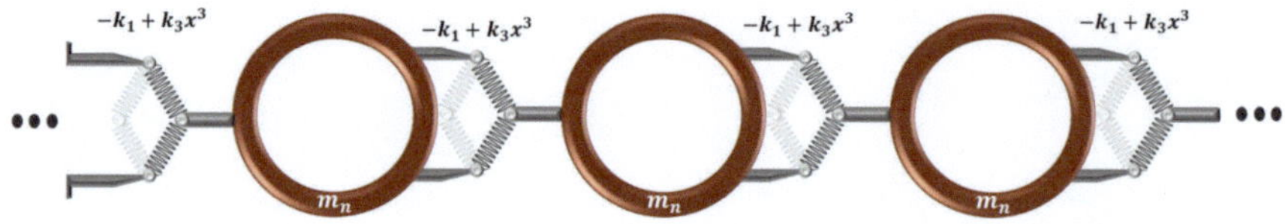

Fig. 6.10 A bistable periodic lattice

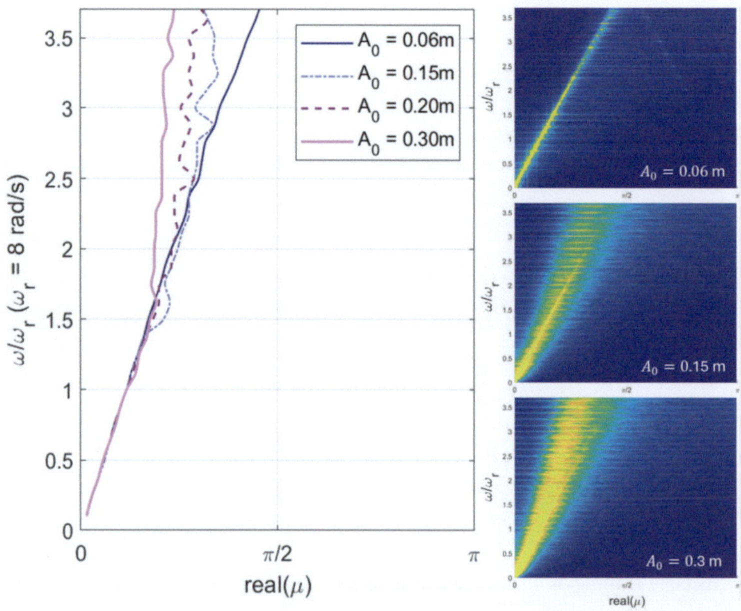

Fig. 6.11 Numerical dispersion bands for a bistable periodic lattice and relevant 2D FFT results

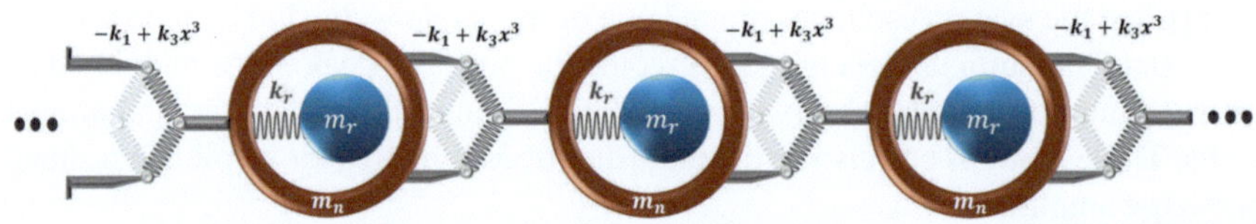

Fig. 6.12 Nonlinear LRM with the bistabity in the main cell springs

resonators, the propagation is continuous through the frequency range. One important aspect is the diffusion of wave propagation peaks, as amplitude increases, similarly to monostable Duffing cases (Fig. 6.11).

Furthermore, the bistablity can be introduced into the main cell springs of the LRM, as presented in Fig. 6.12. Herein, the aim is to extend the harmonic diffusion property of bistable springs to resonators, i.e. diffuse the anti-resonance property to multiple frequencies, thus providing a wider band gap. The numerical analyses

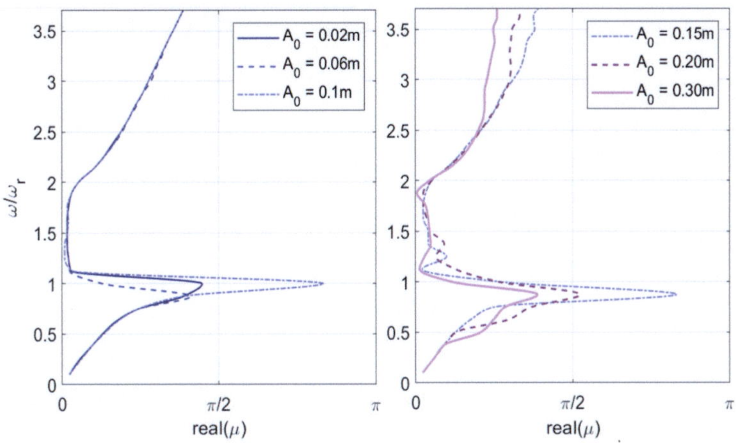

Fig. 6.13 Numerical dispersion results for LRM with bistable main cell springs

were conducted and dispersion relations were obtained for various amplitudes and are presented in Fig. 6.13.

6.4.2 Experimental Validation

Trough the fused deposition modelling (FDM), 3D-printed small-scale unit cells where printed, as illustrated in Fig. 6.14. To realize ultra-low frequency bistable LR finite metamaterials, the unit cells were considered as main cells; the main cell represents the outer mass depicted in Fig. 6.14 and connects to a resonator. Meanwhile, each resonators is only connected to the main frame with flexible linear elastic beams working in simple shear. Main cells connects to other main cells by means of slender beams working in shear too. To exhibit bistability, beams connecting every two main cells were designed to be on the onset of buckling under a preset axial load. The main cells were connected to each other and to the shake table either by linear or bistable beams as depicted in Fig. 6.14. More specifically, the distance between the main cells is 100 mm; then, 100 mm beams for the linear case and 116 mm beams for the bistable case were used. Since beams are considered to bend in shear, slender rectangular beams were considered to prevent yielding.

Apart from beams, all parts were 3D printed with ABS material. This provided sufficient stiffness and durability, as ABS is characterized by high toughness and resistance. In the case of bistable beams, the ABS was preferred since stress localization were not observed. In addition, the TPU material was used to reach lower resonance frequency for the resonator beams. The Cubic infill at 20% was considered for the ABS parts whilst beams were printed with 100% infill.

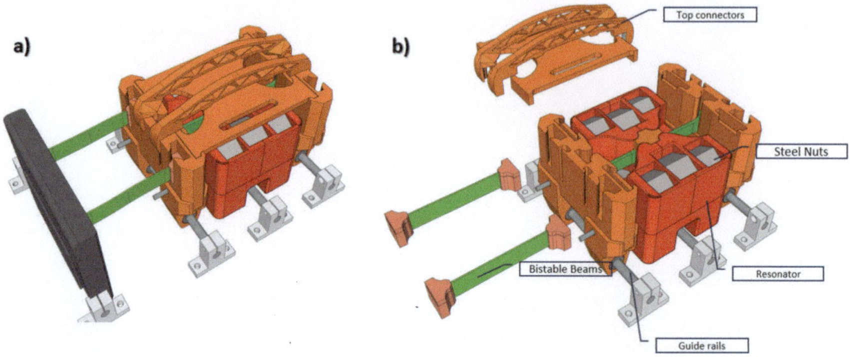

Fig. 6.14 A 3D printed bistable unit cell with the nonlinearity concentrated in the beams: a) the whole system including the resonator; b) the exploded view

The shear stiffness of the beams and the elastic modulus of the materials were determined through displacement-controlled static tests, as depicted in Fig. 6.15. As a result, the elastic modulus of the beams read 1.36 ± 0.15 GPa, whilst the shear stiffness of a linear beam was estimated to be 325 ± 20 N/m. Additionally, the masses of the main cell parts and the resonator were estimated to be 280 and 620 g, respectively. To increase the resonator mass, steel blocks were inserted into pre-designed holes, as indicated in Fig. 6.14b. Accordingly, the frequency of a unit cell without a resonator, was estimated to be 5.9 ± 0.2 Hz, and the resonator frequency was estimated to be 4.5 Hz.

A shake table is illustrated in Fig. 6.16; it is endowed with 500×500 mm and was utilized to provide a dynamic input. It is a TDG Desktop Shake Table v2, and is capable of delivering excitations in the range, 1–2 g, with a payload of 100–50 kg, respectively. It can achieve a maximum velocity of 500 mm/s.

The initial set of tests involved a single linear unit cell with and without a resonator subjected to a frequency sweep excitation, between 1 and 10 Hz. Displacement responses were measured through a laser vibrometer and high speed digital image correlation system. Using the FFT, the frequency content of the response was calculated and compared with the expected frequency responses obtained from accelerometer measurements; see, in this respect, Fig. 6.17. Due to frictional forces, the results revealed a slight increase in the resonant frequencies, i.e. 6.2 Hz.

To identify damping properties including Coulomb and viscous damping, as well as the parameters of the resonator, two types of frequency sweeps were conducted. These sweeps, featuring amplitudes of 5 and 8 mm, respectively, extended up to 10 and 8 Hz. The tests were performed on a single unit cell connected to the shake table using linear beams, both with and without resonator. Consequently, four tests were conducted, during which displacement data from both the shake table and the main frame were recorded. The identification was conducted considering the following system of EoMs:

$$\mathbf{M}\ddot{x} + \mathbf{C}\dot{x} + \mathbf{F}_f(\dot{x}) + \mathbf{F}_k(x) = \mathbf{F}_{ext}(\omega, t) \tag{6.5}$$

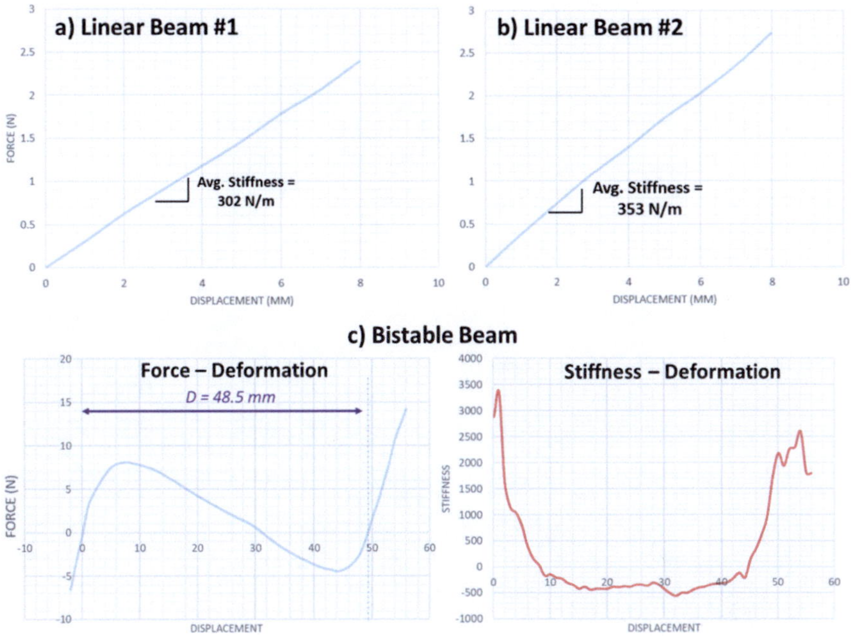

Fig. 6.15 Static identification of linear and nonlinear (bistable beams): **a** Positive stiffness of linear beam #1; **b** Positive stiffness of linear beam #2; and **c** Bistable beam response

where $\mathbf{C} = \mathbf{C}_0 + \mathbf{C}_1\dot{x}$. For a bistable LRM lattice with single cell, the matrices of (6.5) can be expanded as follows:

$$\mathbf{M} = \begin{bmatrix} m_m & 0 \\ 0 & m_r \end{bmatrix} \tag{6.6}$$

$$\mathbf{C}_0 = \begin{bmatrix} c_{m,0} & -c_{r,0} \\ -c_{r,0} & c_{r,0} \end{bmatrix} C_1 = \begin{bmatrix} c_{m,1} & -c_{r,1} \\ -c_{r,1} & c_{r,1} \end{bmatrix} \cdot \dot{x}$$

$$\mathbf{F}_f = \begin{bmatrix} F_{f,m}(x) & -F_{f,r} \\ -F_{f,r} & F_{f,r} \end{bmatrix} \tag{6.7}$$

$$\mathbf{F}_k = \begin{bmatrix} f_k(x) & -k_r \\ -k_r & k_r \end{bmatrix} \tag{6.8}$$

where subscripts m and r represents the main cell and resonator, respectively. The stiffness term, $f_k(x) = k_1x + k_3x^3$, is a cubic polynomial used for the bistable beam, whilst $k_3 = 0$ for a linear beam. For ease of integration, the dry friction term, F_f was expanded following the formulation presented in Specker et al. (2014) and summarised as follows,

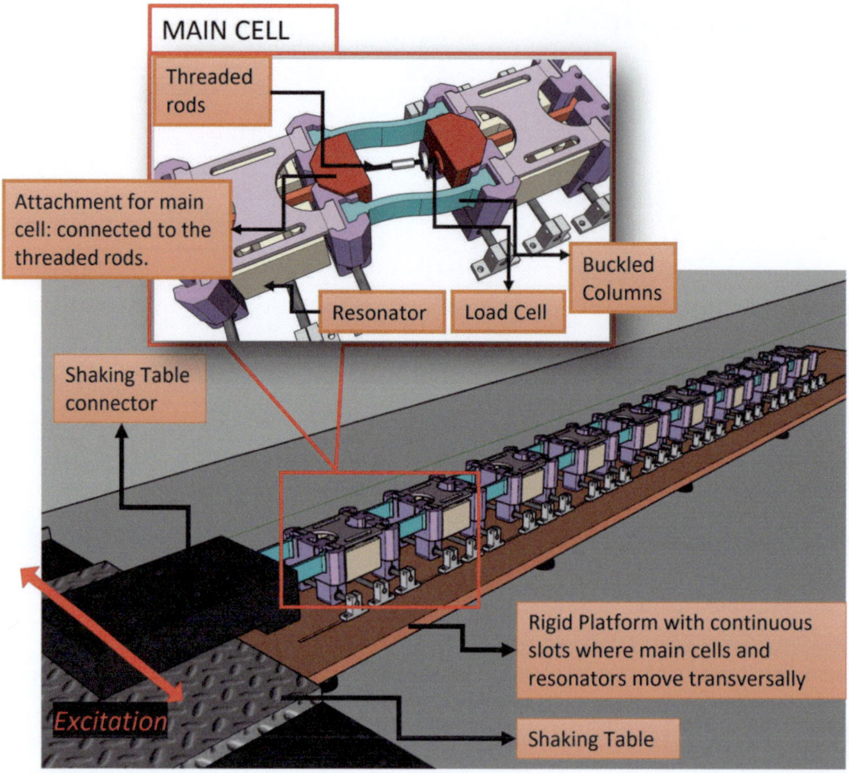

Fig. 6.16 A chain of 3D FFF printed nine bistable unit cells connected to the TDG Desktop Shake Table v2

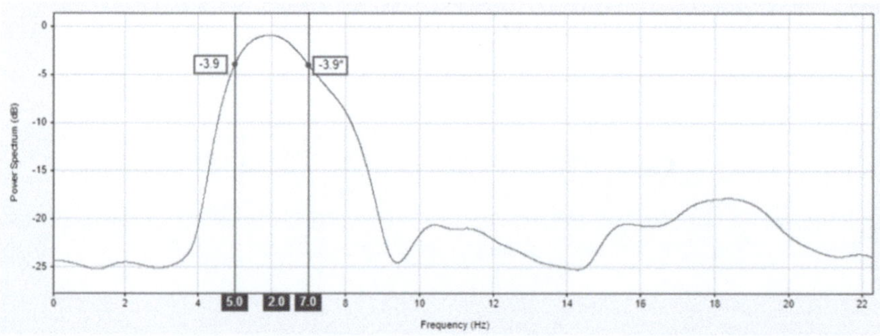

Fig. 6.17 FFT of an accelerogram obtained from MPU6050 on the top connector; see Fig. 6.14

$$F_f = F_{st} + F_{kin}\left(\tanh(\dot{x}/\dot{x}_{vt}) - g_f\right) \tag{6.9}$$

$$g_f = \frac{\dot{x}}{\dot{x}_{sp}} e^{\left(0.5-\left(\frac{\dot{x}}{\sqrt{2}\dot{x}_{sp}}\right)^2\right)} \tag{6.10}$$

where F_{st} and F_{kin} define the static and kinetic friction terms. Accordingly, the proposed numerical model requires a total of 13 parameter to be identified: (i) stiffnesses, k_1, k_3, k_r ; (ii) viscous damping constants, $c_{m,0}, c_{m,1}, c_{r,0}, c_{r,0}$; (iii) static and kinetic friction forces, $F_{s,m}$, $F_{k,m}$, $F_{s,r}$, $F_{k,r}$; and (iv) the Stribeck peak, and the kinetic friction velocity terms, $\dot{x}_{sp}, \dot{x}_{vt}$. To determine these 12 parameters, the numerical model of the unit cell was subjected to a driven boundary input, utilizing input histories recorded from shake table tags. By applying the Newmark method, the numerical response of the unit cell was obtained. The parameters were then determined through the minimization of the experimental FRFs and numerical FRFs, within a multi-variate optimization problem solved by a Genetic Algorithm. The optimized values, i.e. identified parameters, are collected in Table 6.2, whilst the FRFs comparisons are depicted in Figs. 6.18 and 6.19.

Table 6.2 Identified parameters

	k_1 (N/m)	k_3 (N/m^3)	c_0 (Ns/m)	c_1 Ns^2/m^2	$F_{f,s}$ (N)	$F_{f,k}$ (N)
L. Mc.*	441.8	–	1.30	6.7	1.4	0.76
B. Mc.*	1135.7	2.15e6	1.30	6.7	1.4	0.76
Res.*	552.3	–	0.96	14.9	9.2	0.45

* L.Mc.: Linear Main Cell, B.Mc.: Bistable Main Cell, Res.: Resonator

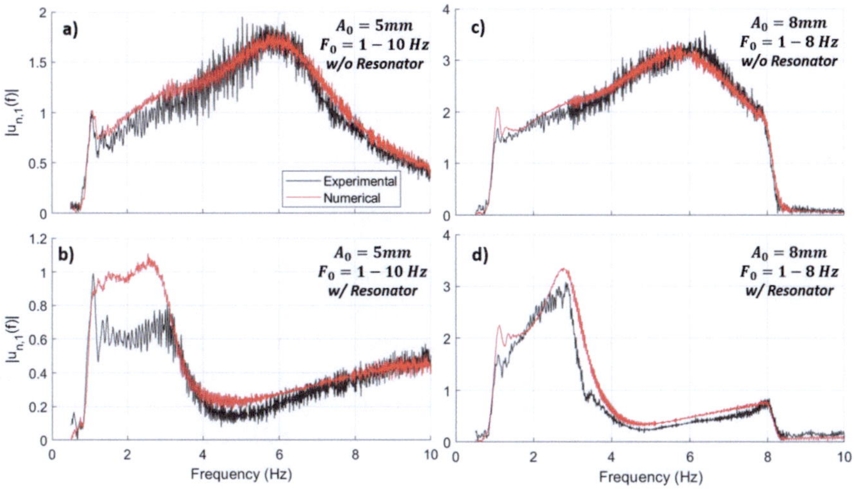

Fig. 6.18 The FRFs comparison between numerical and experimental results evaluated on a linear unit cell

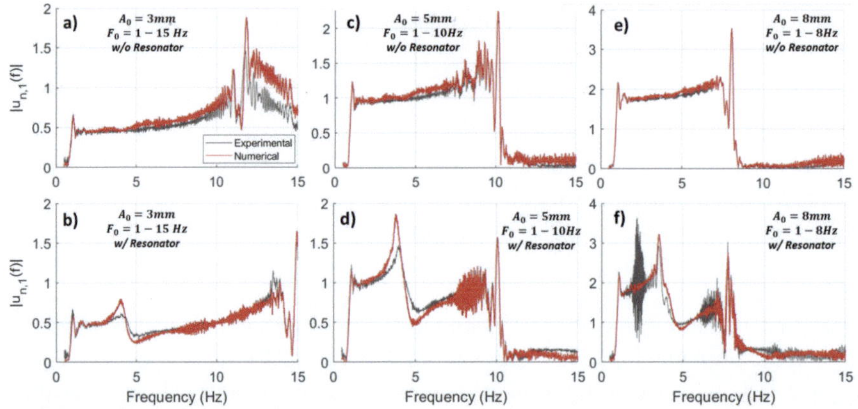

Fig. 6.19 The FRF comparison between numerical and experimental results for bistable unit cells

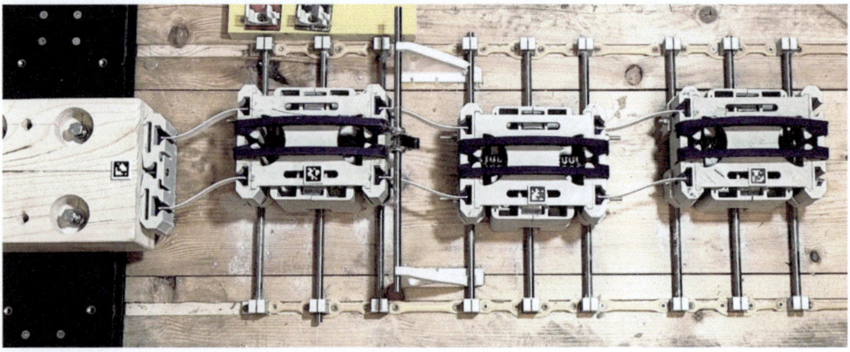

Fig. 6.20 Finite lattice with 3 bistable unit cells

Following the primary objective of bistable LRMs, the subsequent results present experimental findings regarding wave attenuation characteristics. In essence, chains consisting of single and three linear and bistable unit cells were subjected to both frequency sweep and sinusoidal inputs. The setup for the three-cell configuration is illustrated in Fig. 6.20.

Displacement time history results of single linear and bistable unit cells subjected to frequency sweep are presented in Figs. 6.21 and 6.22, respectively. Unit cells were tested both with and without resonator attached. The effect of the resonator can be seen in both cases where significant wave attenuations are achieved.

Since bistable oscillators have an offset stable point, a better comparison can be made in the velocity domain. Accordingly, the velocity FRFs are shown in Fig. 6.23. The presented cases show the benefits of resonators rather than a direct performance comparison between linear and bistable cases. Since bistable LRM exhibit wide range of dynamics, a more comprehensive investigation is necessary for direct comparison between linear and bistable cases.

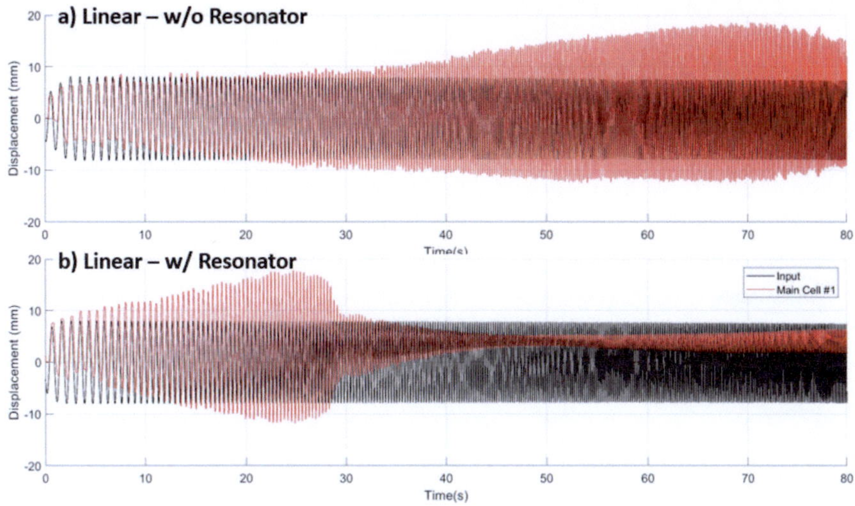

Fig. 6.21 Single linear unit cells with and without resonator subjected to a frequency sweep (8 mm, 1–8 Hz)

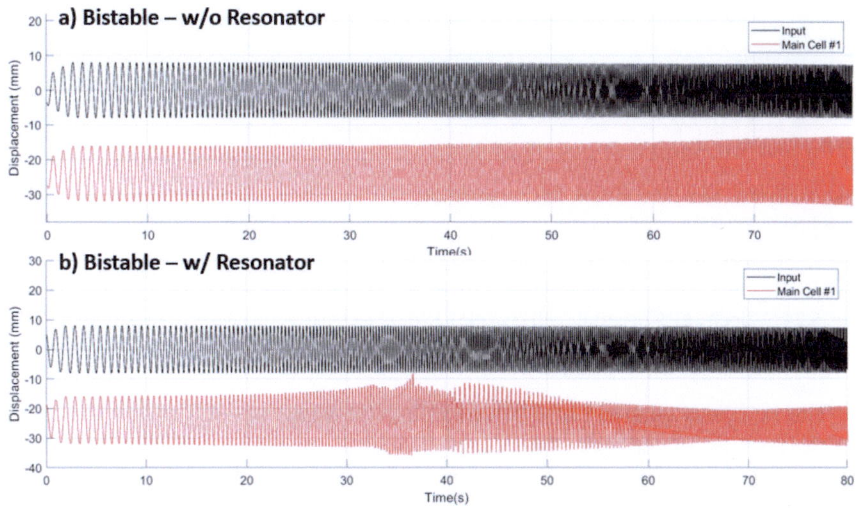

Fig. 6.22 Single bistable unit cell with and without resonator subjected to a frequency sweep (8 mm, 1–8 Hz)

Considering the triple unit cell setup, provided in Fig. 6.20, the velocity FRFs considering a frequency sweep at 8 mm amplitude are presented in Fig. 6.24 for the bistable LRMs. When a resonator is attached and despite the fact that it is not tuned for high frequencies, it can be observed that the a bistable LRM prevents the high frequency content to be transmitted to the last cell of the chain.

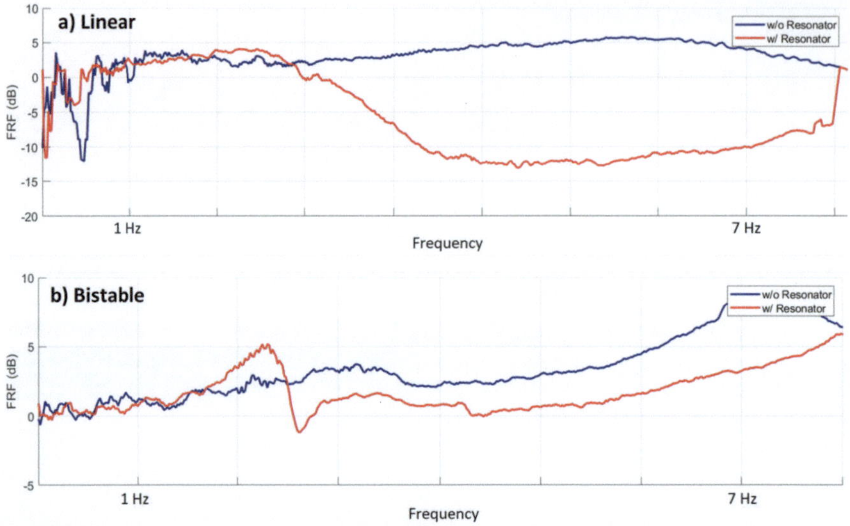

Fig. 6.23 Velocity FRF of Single unit cell with and without resonator subjected to frequency sweep (8 mm, 1–8 Hz)

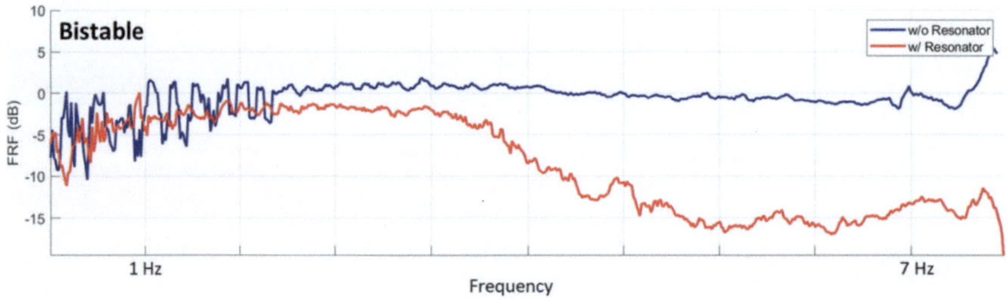

Fig. 6.24 Velocity FRF of a triple cell bistable LRM subjected to a frequency sweep (8 mm, 1–8 Hz)

The experimental study has yielded valuable insights into the effects of friction on the response. It has become evident that for a precise modeling of real-world applications with low-frequency elastodynamic metamaterials, the influence of friction, particularly static friction, must be taken into account. Furthermore, it has been demonstrated that damping resulting from 3D printed parts, necessitates high-order approximation modeling, as the linear model with viscous damping resulted to be inadequate.

6.5 Linear and Bistable Metafoundations

6.5.1 *Basic Aspects of a Metafoundation*

As underlined in the Sect. 6.1, locally resonant metamaterials (LRM) are commonly considered in the low-frequency regime, due to their ability to exhibit low-frequency bandgaps. The fundamental concept involves designing a periodic structure that possesses stop bands capable of prohibiting or attenuating the propagation of elastic waves within a desired frequency range. In particular, the metafoundation (MF) comprises a finite number of unit cells realized with a main frame and an internal moving reinforced concrete (RC) mass, that acts as a resonator. The main cell consists of two primary components: flexible columns and an RC slab, as depicted in Fig. 6.25b. The main cell serves two essential purposes: first, it must possess high flexibility to increase the structure's period, and second, it should have sufficient strength to support the substantial weight of the resonator blocks. The resonators are connected to the main frame slab through springs, as illustrated in Fig. 6.25b. There are various approaches for realizing these springs; however, wire ropes prove to be a suitable choice due to their three-dimensional flexibility and high damping capabilities (Basone et al. 2019b).

To demonstrate the performance, parameters, and characteristics of the MF, a slender petrochemical tank was chosen as a case study, as depicted in Fig. 6.26a. The tank compromises a radius of 8 and 24 m liquid height. Accordingly, the simplified

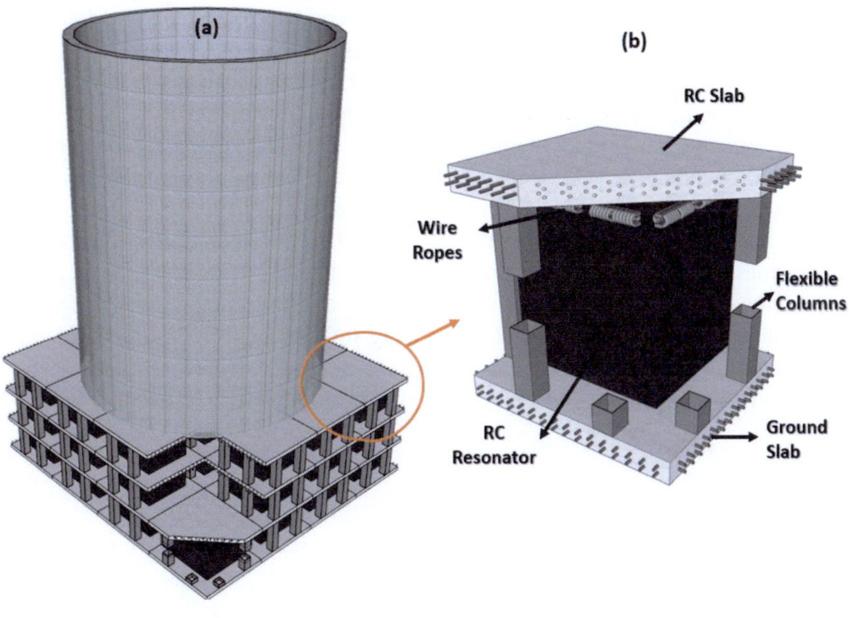

Fig. 6.25 **a** Triple layer metafoundation layout for slender storage tank; **b** Unit cell details

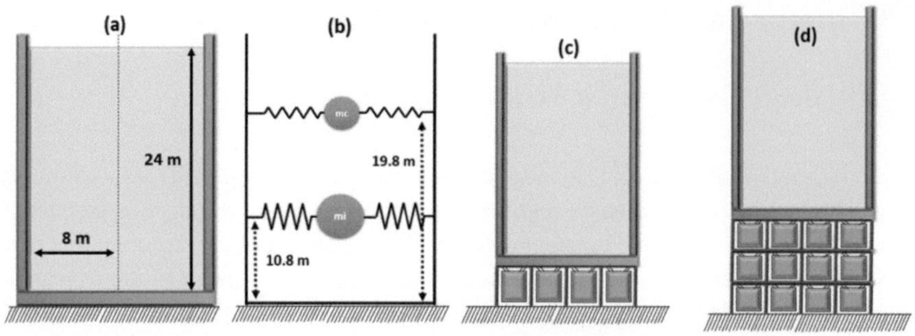

Fig. 6.26 Details of the superstructure and the MFs: **a** Slender petrochemical tank; **b** Simplified dynamical modelling of the tank; **c** Tank with a single layer MF; **d** Tank with a triple layer MF

Table 6.3 Details of the simplified tank model

Impulsive			Convective		
m_i	4014.8	Tons	m_c	762.4	Tons
c_i	5790.6	kNs/m	c_c	11.4	kNs/m
k_i	835184.1	kN/m	k_c	1717.7	kN/m

hydrodynamic response model suggested by Malhotra et al. (2000) that provides the correct dynamical superstructure response on the MF due to impulsive and convective masses is implemented. The properties of the simplified model are provided in Table 6.3 and presented in Fig. 6.26b. It can be seen that due to the slenderness of the tank, impulsive mass contributes a significantly larger mass ratio; accordingly, throughout the research, the results are given for the impulsive mass; nevertheless, the first convective mode is still included in the modelling and analysis stages.

To assess and standardize the performance of different MFs, a set of artificially generated ground motion (GM) records was utilized. These motions were generated in agreement with the Eurocode 8 (EC8) (European Committee for Standardization 2004b) response spectra (Mitseas and Beer 2019). The evaluation was conducted at locations with design peak ground accelerations (PGA) of 0.3, 0.5, and 0.7 g, with all locations assumed to have soil type B. To ensure a comprehensive analysis, time history simulations were carried out using 100 artificially generated horizontal accelerograms per location. The Newmark integration scheme—constant average acceleration—was employed for accurate numerical integration of EoMs (Fig. 6.27).

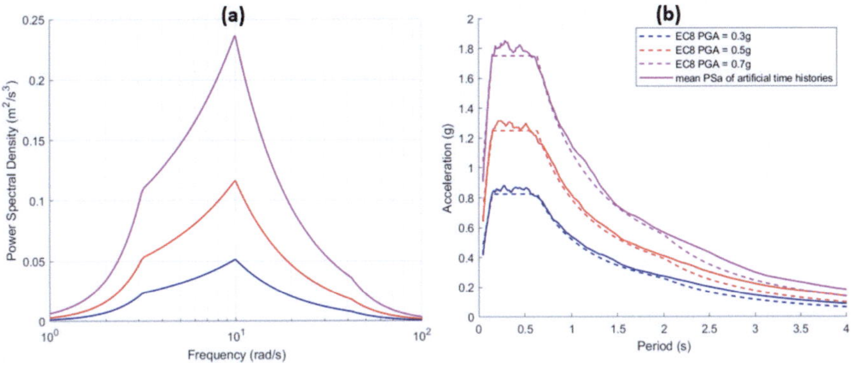

Fig. 6.27 **a** Single sided power spectral density of design spectra; **b** comparison of design spectra against mean spectra of artificially generated GMs

6.6 Linear Metafoundation

The MF combines the effects of conventional seismic isolation, tuned mass dampers, and low-frequency bandgaps effects of the LRM. Consequently, the performance of the MF is influenced by multiple parameters, which are thoroughly investigated and explained in the following sections.

6.6.1 Main Cell Parameters

To achieve maximum isolation effect, the columns of the main cell should be designed using the most flexible section that satisfies the design criteria. Several factors such as section type, material, number of columns (N_{col}), and column height (H_{col}) in a layer can influence the stiffness. The selection of section type and material depends on factors like availability, requirements, and the engineer's design choices. The number of columns can be determined based on the number of unit cells considered in a layer. Referring to Fig. 6.25b, the number of cells in the -X direction (N_{Cx}) and the -Y direction (N_{Cy}) affect theN_{col}. To understand the impact of N_{col}, multiple single-layer MFs were designed with different N_{Cx} values (assuming $N_{Cy} = N_{Cx}$). Figure 6.28a presents the variations in horizontal and vertical stiffness. It is observed that increasing N_{Cx} and N_{Cy} leads to lower stiffness in the horizontal direction but results in a stiffness increase in the vertical direction. Consequently, the determination of the number of cells creates a Pareto front optimization problem, which requires a trade-off between horizontal and vertical stiffness. The Pareto front plot is shown in Fig. 6.28b. The Pareto front represents the optimal horizontal and vertical stiffness values and the N_{Cx} can be selected either by a decision-maker considering additional factors or through methods such as the sum of weighted sums or the utopian point, as depicted in Fig. 6.28b.

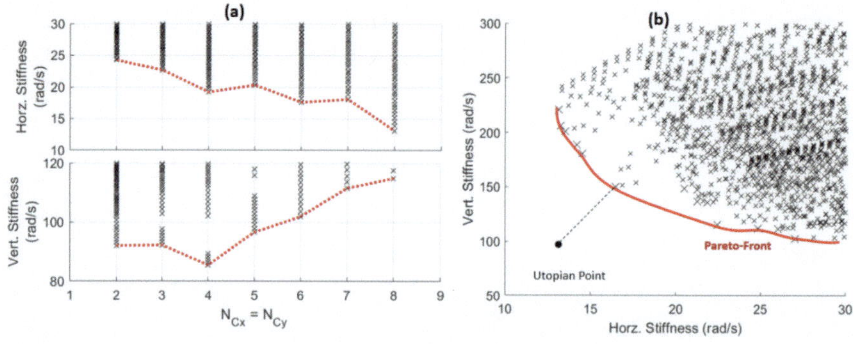

Fig. 6.28 Design results with various N_{Cx} ($= N_{Cy}$) values: **a** Variation of horizontal and vertical stiffness of the structure wrt N_{Cx} ($= N_{Cy}$); **b** Pareto front representation of horizontal and vertical stiffness values together with the optimum solution wrt the utopian point

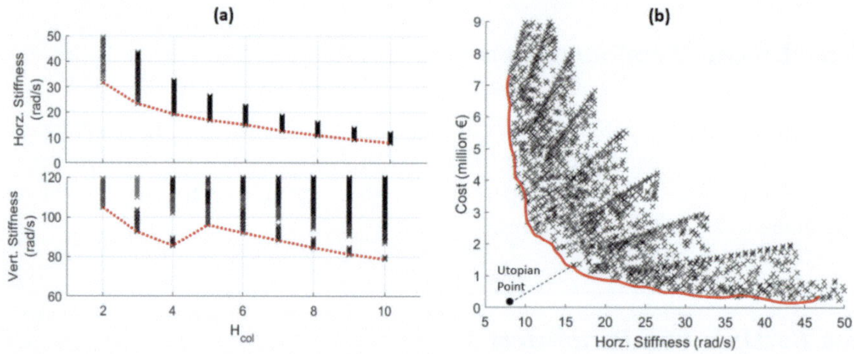

Fig. 6.29 Design results with various H_{col} values: **a** Variation of horizontal and vertical stiffness of the structure wrt H_{col}; **b** Pareto front representation of horizontal and vertical stiffness values together with the optimum solution wrt the utopian point

The height of the columns directly impacts the flexibility in both the horizontal and vertical directions. This aspect of the MF offers significant advantages, as the performance can be enhanced by adjusting the H_{col}. However, it is important to note that, as demonstrated by Guner et al. (2022), that due to the additional material and excavation costs involved, the cost of the MF rapidly increases as H_{col} increases. For economic viability, the cost of a single-layer MF and the corresponding horizontal flexibility for various column heights are presented in Fig. 6.29. Once again, it is evident that there is a trade-off between cost and flexibility, resulting in a Pareto front problem that requires either a decision or a logical algorithmic selection.

6.6.2 Resonator Parameters

In the case of a single-layer MF, the resonators act as tuned mass dampers for uncoupled translational degrees of freedom. Therefore, a vast body of background and parametric research on tuned mass dampers can readily be applied to the resonators of the MF. One crucial factor in designing the resonators is the determination of the resonator mass. Based on knowledge about tuned mass dampers, it is well-established that an increase of the resonator mass corresponds to an improve in performance. Hence, it is advisable to utilize the maximum mass of the resonator. It is worth noting that the size of the resonators depends on H_{col}, N_{Cx} and N_{Cy} assuming that the foundation size is fixed and determined by the dimensions of the superstructure. Therefore, there is a clear linear relationship between H_{col} and the total resonator mass in a layer (M_{Res}). Conversely, N_{Cx} and N_{Cy} have a reverse and nonlinear relationship, as an increase in the number of cells leads to an increase in N_{col}, reducing the available space for the resonators; refer to Fig. 6.30a, in this respect. Additionally, since each resonator requires a certain distance from the surrounding columns (see Fig. 6.25a), the decrease in M_{Res} increases as N_{Cx} and N_{Cy} grow. Together with the previous flexibility objectives, the 3D optimization problem can be shown as depicted in Fig. 6.30b. It can be seen that the utopian point exhibits the lowest stiffness values with the maximum resonator mass. For the current case, the closest point was determined at $N_{Cx} = N_{Cy} = 4$ and considered in the rest of the study.

The effectiveness of tuned mass dampers relies on properly tuning the resonator stiffness and damping parameters in relation to the characteristics of the superstructure. Numerous options and intelligent optimization procedures are available in the literature, although they are not elaborated upon in this context. In this research, a frequency-based H_∞ optimization procedure has been employed, which takes into

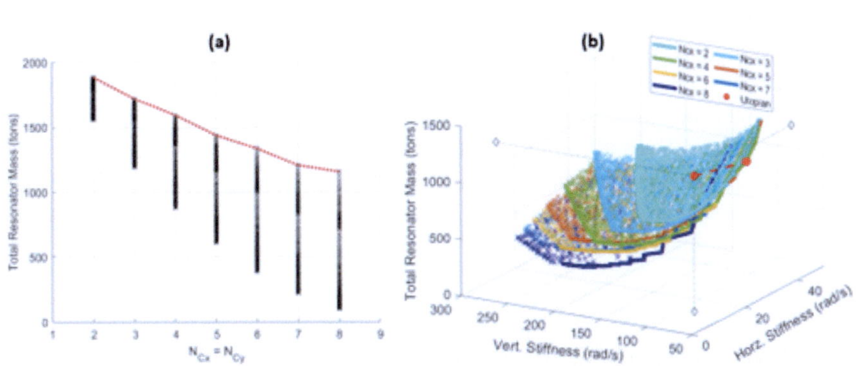

Fig. 6.30 Design results with various N_{Cx} ($= N_{Cy}$) values: **a** Variation of the total resonator mass in the MF wrt N_{Cx} ($= N_{Cy}$); **b** Pareto front representation of horizontal stiffness, vertical stiffness and total resonator mass values together with the optimum solution wrt to the utopian point. The optimum was located at $N_{Cx} = N_{Cy} = 4$

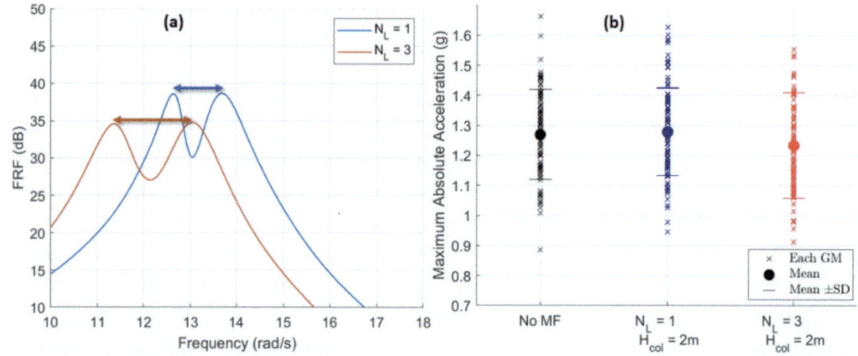

Fig. 6.31 Comparison of 2 and 6 m MFs with $N_L = 1$ and $N_L = 3$: **a** Frequency response functions behaviour; **b** Time history analysis results

account the site soil conditions through an input power spectral density (PSD) function (Guner et al. 2022). On the other hand, for nonlinear cases, where the system linearization is not feasible, the relevant optimization can be conducted in the time domain using methods such as design point selection (e.g., central composite design, Latin hypercube sampling) and surface fitting techniques like Kriging (Guner et al. 2023).

It is crucial to highlight that the MF incorporates multiple resonators, which can act in parallel, series, or complex coupling arrangements. This characteristic allows for the creation of wide ranges of attenuation due to locally resonant effects. Since earthquakes typically have a broad frequency content, a wider range of attenuation provides better protection and robustness against randomness. MFs with $N_L = 1$ and $N_L = 3$ were designed with the same $H_{col} = 2\,m$ and both frequency response functions and time history analyses results are presented in Fig. 6.31. With an increasing number of layers, the attenuation around the resonance frequency also increases, in terms of both amplitude and frequency range. However, since H_{col} was kept constant, the MF height increases. A similar investigation has been made by keeping the MF height the same at 6 m and considering two MFs with $N_L = 1$ and $N_L = 3$ ($H_{col} = 6$ and 2 m, respectively). The relevant results are presented in Fig. 6.32: one can observe that the increased layer stiffness in $N_L = 3$ results in a worst performance than the single-layer case.

6.7 Bistable Metafoundations

Although bistability was considered earlier for seismic isolation (Plaut et al. 2008; Jeffers et al. 2008), the protection focussed on the vertical component of seismic waves. In this respect, linear metafoundations have also been profitably proposed to mitigate the vertical motion of storage tanks and small modular reactors (Franchini

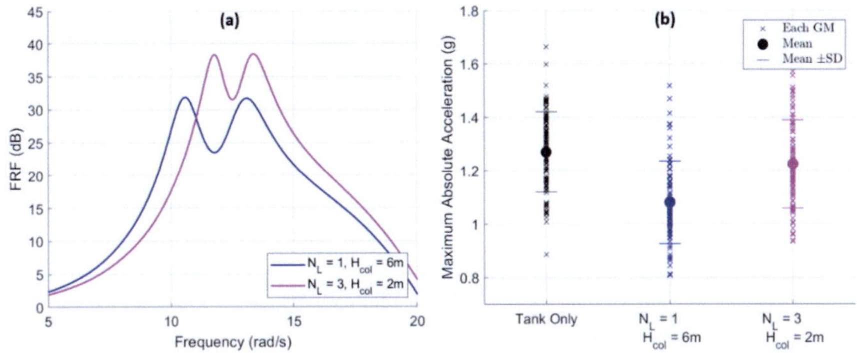

Fig. 6.32 Comparison of 6 m MFs with $N_L = 1$ and $N_L = 3$: **a** Frequency response functions; **b** Time history analyses results

et al. 2020; Guner et al. 2022). Conversely, the concept of combining bistability with Locally Resonant Metafoundations (MFs) devoted to the seismic isolation of process components against the horizontal component of waves is largely unexplored. To address this missing issue, four different MFs were designed with single and double layer configurations as well as endowed with linear and bistable columns. For clarity, the single-layer and the double-layer MFs are identified with the prefix 1L- and 2L-, respectively. Along this line, the structural details of the 2L-BM are depicted in Fig. 6.33. More precisely, each metafoundation cell consists of two parts: (i) the primary cell is composed of four flexible slender steel columns and two concrete slabs; (ii) the concrete resonator that is connected to the slab with steel wire ropes. The relevant design followed both Eurocode 1993 Part 1-1 (European Committee for Standardization 2005) and Eurocode 1998 Part 1 (European Committee for Standardization 2004a). The construction site was selected in Priolo Gargallo in Sicily, Italy, and is characterized by soil type B with peak ground acceleration (PGA) equal to 0.56 g for a return period of 2475 years. Each layer has a height H = 1 m and contains 9 unit cells in a 3 by 3 layout. The superstructure is a slender fuel unanchored storage tank that is a part of an existing plant (Salandra et al. 2017; Carta et al. 2016).

To introduce the bistability effect into the MFs, pre-buckled steel columns were conceived. In particular, columns were designed to elastically buckle due to gravity loading along their weak axis. The weight of the tank was assumed to be constant at its maximum capacity for the safe life limit state. The foundation is symmetric in the x and y axes; as a result, half of the columns are designed to buckle on the x-axis whilst the rest buckle on the y-axis. One of the critical points of the buckled columns is that once the buckling load is reached, the stable region where the column stiffness in the lateral direction is positive is comparatively small and sensitive to axial loading. To overcome this issue, the strong axis of the buckled columns was used as a restraining mechanism for lateral displacements of buckled columns. However, it is not favourable to add the strong axis stiffness of columns to the stiffness of

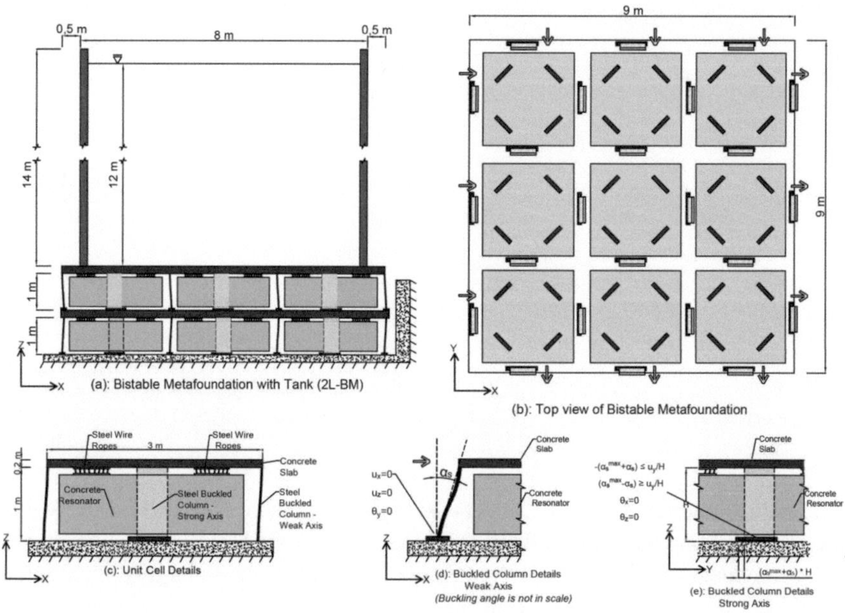

Fig. 6.33 **a** A bistable MF with a storage tank; **b** Top view of the bistable LRMF; **c** Unit cell details; **d** Weak axis view of a buckled column; **e** Strong axis view of a buckled column

the snap-through region. Consequently, the column ends are designed such that they allow unrestrained motion in the strong axis up to an amount of allowed snap-through buckling distance. In other words, between two stable points of weak axis buckled columns, there is no additional stiffness induced from the columns in the strong axis. However, in the case of a significant demand, the strong axis of the columns contributes to the lateral resistance and prevents instability.

Due to the coupled mechanism of buckling where the lateral stiffness is bound with axial load, the design of the columns becomes a challenging problem. In the design process, two main points were considered. In the weak axis, the columns should be weak enough and slender, with a slenderness ratio $L_e/r = 200$, to elastically buckle due to the superstructure load; but they should exhibit a high moment capacity, to remain elastic after buckling. Furthermore, in the strong axis, they must resist lateral loads. The selection of the steel column section was governed by weak axis limitations. In this respect, two parameters are important: the section width b and the flange thickness t_f; to minimize the buckling load, the weak axis moment of inertia must be reduced and depends on b^3 and t_f; to maximize the moment capacity, b^2 and t_f should be increased. As the effect of b is more prominent on the buckling load, t_f should increase to improve the moment capacity. Therefore, t_f is maximized, and the favourable section is the rectangular one.

Preliminary analyses carried out with rectangular sections, entailed that the allowed post-buckled drift angle α_s of the column, significantly influences the system response. The α_s value can be also defined as the drift angle of snap-through buckling range in both directions $(-\alpha_s, \alpha_s)$ or the drift angle that corresponds to half of the distance between stable points and reads,

$$\alpha_s = \arctan\left(\frac{u_{top}^{pb} - u_{bottom}^{pb}}{H}\right) \tag{6.11}$$

where u^{pb} represents the post-buckled displacement and H defines the storey height. This angle can be adjusted during the design stage by limiting the drift capability of the column using the strong axis allowance of the adjacent columns.

To present the benefits of the bistable MFs (BMs), also linear MFs (LMs) were designed and analyzed. The corresponding columns were designed to remain elastic and undamaged for the considered seismic design level; therefore, the hollow square steel sections were considered with a S355 steel grade. The dimensions of the designed cross sections are gathered in Table 6.4.

To achieve the best performance from each BM, an optimization process on some mechanical parameters is deemed necessary. The connection of the resonators to the primary cell is realized by means of linear springs and dashpots can be subject to optimization of the MFs. For each resonator mass, in view of efficiency, the largest mass compatible with the unit cell dimensions is taken (Reggio and Angelis 2015; Basone et al. 2019a). In addition to the resonator variables, the allowed snap-through buckling range $[0, \alpha_S^{max}]$ is also considered as an optimization parameter, where α_S^{max} is computed from the elastic limits of the column steel section. In fact, low α_s values trigger more snap-through motions, and this behaviour entails wanted energy dissipation. The performance index (PI) adopted in the optimization of the BM is the energy dissipation index (*EDI*), Basone et al. (2019a), that reads,

$$EDI = \frac{\sum E_d^{col} + \sum E_d^{res}}{\sum E_d^{col} + \sum E_d^{res} + E_d^{imp} + E_d^{con}} = \frac{\sum E_d^{col} + \sum E_d^{res}}{\sum E_d} \tag{6.12}$$

In particular, the EDI is defined as the ratio of dissipated energy by the LRMF with respect to the whole structure. Since all elements of the metastructure are elastic, and the bistable system exhibits no hysteretic dissipation, the energy can only

Table 6.4 Design column dimensions and parameters

LM	Layer	Width (mm)	Thickness (mm)	BM	Layer	Width (mm)	Depth (mm)	α_S^{max} deg
1L	1	130	30	1L	1	360	16	3
2L	1	160	30	2L	1	425	16	3
	2	150	30		2	385	16	3.3

be dissipated through viscous damping. The details of the optimization are available in Guner et al. (2023). To validate the optimized solutions, nonlinear time history analyses were carried out. For completeness, earthquakes associated with the design basis earthquake (DBE) and the safe shutdown earthquake (SSE) events were considered; moreover, the resonators were characterized by the optimized mechanical parameters.

In view of completeness, in addition to the EDI defined in Eq. (6.11), the seismic input energy provided to the superstructure E_{SS} is evaluated. E_{SS} can be computed as follows,

$$E_{SS} = m_i \cdot \int_0^{\tau} a_g(t) \cdot v_i(t) \cdot dt + m_c \cdot \int_0^{\tau} a_g(t) \cdot v_c(t) \cdot dt \tag{6.13}$$

where, m and v(t) define both mass and velocity whilst subscripts i and c identify impulsive and convective masses, respectively; $a_g\ (t)$, instead, defines the ground acceleration at time t, whilst τ defines the generic duration of each seismic record. Clearly, an efficient performance of the metafoundation limits the amount of E_{SS}.

Therefore, for each DBE and SSE limit state, eight seismic records were selected to assess the performance of the proposed LRMFs, respectively. The selected seismic records are collected in Table 6.5 whilst the acceleration spectra, their mean, and mean plus one standard variation are plotted in Fig. 6.34.

Herein, Fig. 6.35 depicts the EDI values for each seismic record of Table 6.5 both for the 1L and 2L LRMFs, respectively, with their median values. In both cases, the BM entails more than 10% increased median EDI values whilst keeping EDI value above 70% for all events. Clearly, there are a couple of SSE events where the 1L LM does not exhibit a favourable behaviour. Conversely, the 2L BM exhibit a favourable median EDI for both DBE and SSE events. Consequently, the bistable snap-through motions in the metafoundations cause an increased damping and prevent damage to the superstructure.

The advantages of the proposed bistable metastructures can be clearly appreciated through the estimate of E_{SS}. The results are depicted in Fig. 6.36 for both DBE and SSE events, respectively. In the case of a single layer, see Fig. 6.36a and c, given the large amounts of E_{SS}, the BM underperforms with respect to the LM in almost all cases. Though the 1L-BM is characterized by a favourable EDI, the energy dissipated in the BM is not enough. The 2L-BM instead, entails a different isolation effect due to the presence of 2 layers of bistable columns. More precisely, in the case of DBE events, the 2L-BM transfers only 6.1% of energy to the superstructure, in terms of median, compared to the 2L-LM case. With regard to the SSE events, see Fig. 6.36d, reductions of about 8.9% of E_{SS} in terms of median are achieved. Clearly, there are cases, the 1st and 3rd SSE event, where the 2L-BM underperformed with the high E_{SS}. The seismic records and their frequency content causes the 2L-BM to work outside of the snap-through region delimited by α_s. Therefore, the strong axis of the columns significantly contributes to the response, with an increase in stiffness and high energy transfer E_{SS}.

Table 6.5 List of selected seismic records and relevant parameters

DBE events	GM no.	Location	Country	Date	MW	Rjb (km)	PGA (g)
	1	Ano Liosia	Greece	07/09/1999	6	14	0.31
	2	South Iceland	Iceland	17/06/2000	6.5	15	0.48
	3	South I. (aftershock)	Iceland	21/06/2000	6.4	12	0.39
	4	L'Aquila Mainshock	Italy	06/04/2009	6.3	5	0.40
	5	L'Aquila Mainshock	Italy	06/04/2009	6.3	4	0.45
	6	L'Aquila Mainshock	Italy	06/04/2009	6.3	6	0.33
	7	L'Aquila Mainshock	Italy	06/04/2009	6.3	5	0.66
	8	Northridge-01	USA	17/01/1994	6.7	35	0.32
SSE events	GM no.	Location	Country	Date	MW	Rjb (km)	PGA (g)
	1	Erzincan	Turkey	13/03/1992	6.6	13.0	0.39
	2	South Iceland	Iceland	17/06/2000	6.5	7.0	0.63
	3	South Iceland	Iceland	21/06/2000	6.4	11.0	0.42
	4	L'Aquila	Italy	06/04/2009	6.3	4.6	0.44
	5	L'Aquila	Italy	06/04/2009	6.3	4.4	0.49
	6	L'Aquila	Italy	06/04/2009	6.3	4.9	0.55
	7	Landers	USA	28/06/1992	7.3	11.0	0.27
	8	Northridge-01	USA	17/01/1994	6.7	20.1	0.57

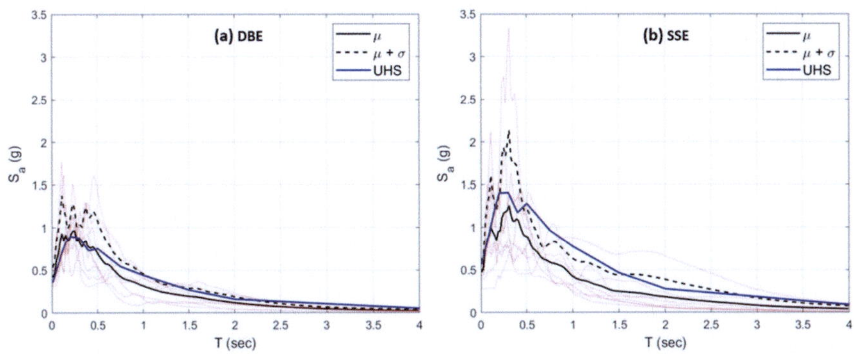

Fig. 6.34 Response spectra of the selected seismic records: **a** DBE limit state; **b** SSE limit state

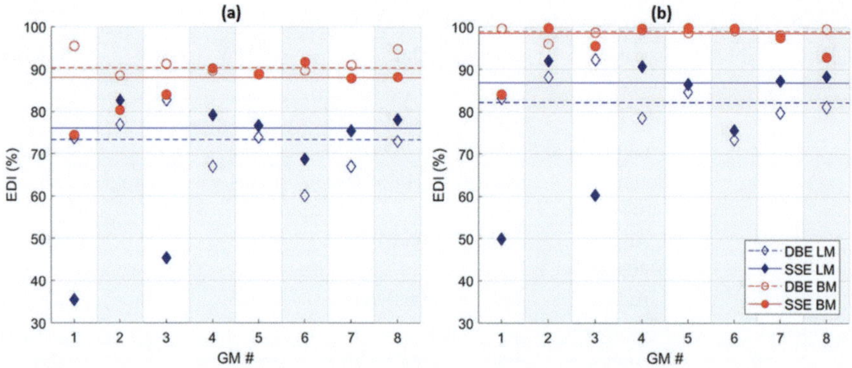

Fig. 6.35 EDI results for each seismic record collected in Table 6.5, where the solid and dashed lines represent the median for SSE and DBE events, respectively: **a** 1L LRMFs case; **b** 2L LRMFs case

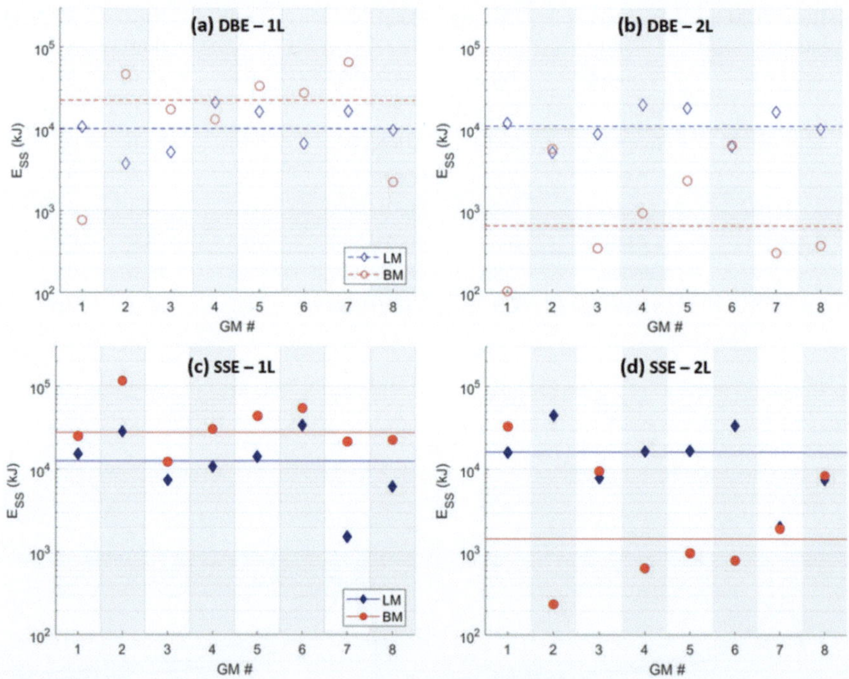

Fig. 6.36 Plots of the energy E_{SS} transferred to the superstructure for the seismic records of 6.5. Continuous lines define median values: **a** DBE-1L case; **b** DBE-2L case; **c** SSE-1L case; **d** SSE-2L case

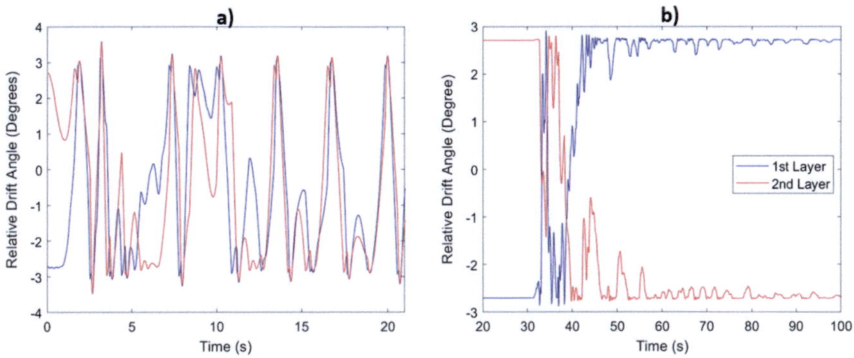

Fig. 6.37 Time history relative drift angle responses of the 2L-BM main slabs: **a** SSE GM-1 record; **b** SSE GM-5 record

The time history displacement responses of the main slabs relevant to the 2L-BM in terms of relative drift angle are depicted in Fig. 6.37 for the GM-1 and the GM-5 SSE seismic records—1992 Erzincan and 2009 L'Aquila events of Table 6.5. The first event entails the worst case for the 2L-BM and the 2L-LM performs better. In particular, from Fig. 6.37a, one can argue that the first layer switches from the starting position to the other stable point and then, both layers move in a synchronized manner. Therefore, both layers act together rather than creating asymmetric modes. In particular, the GM-1 record excites low-frequency modes where both layers move together and the energy transfer to the superstructure increases. The opposite behaviour can be observed in Fig. 6.37b, where both layers vibrate and even columns buckle from one stable point to the other one, simultaneously, in opposite directions. This type of motion shows that the movement of the first layer slab governs and, simultaneously, excites the connected resonators with significant energy dissipation.

Time history analyses conducted on optimized bistable metafoundations (BMs) highlighted the favourable behaviour of these isolation systems in terms of reduction of the total seismic energy input transferred to the superstructure. Clearly, the nonlinearity that characterizes the bistable mechanism significantly reduces the total structure stiffness and results in more compact locally resonant metafoundations wrt to the linear metafoundations.

6.8 Conclusions and Future Perspectives

6.8.1 Conclusions

An innovative methodology for the assessment of state-dependent fragility functions has been proposed. It is based on the combination of advanced surrogate modelling techniques, like bootstrapped-PCE, with data from FE models or real world to yield state-dependent fragilities. These fragility curves are functions of the selected IM, and the initial damage condition limit state. The proposed method has been applied to a vertical tank of a more realistic industrial substructure archetype, comprehensively tested by means of a shaking table. The application to a complex and realistic industrial case study has allowed for the derivation of state-dependent fragility curves for a component that proved to be critical during the experimental campaign: the vertical tank mounted on the first floor of the 3-storey braced frame industrial steel structure. Through the estimation of the efficiency index, β_{eff}, it was possible to identify, among the optimal IMs, the relevant occurrence of the E ASAR67. This IM is of particular interest, since it considers the damage experienced by the system through a frequency range shift. Successively, empirical fragility curves were derived by leveraging the large number of outputs generated from the bootstrap and PCE combination. The best probability distribution was then inferred by employing the maximum likelihood estimation method and the Akaike information criterion.

With the aim of improving the performance of locally resonant metafoundations, bistability-based nonlinearities were introduced in the primary cells. The bistability was governed by elastically buckled steel columns, that can freely snap-through from one stable point to another. Along this line, an optimization framework based on the Central Composite Design and Kriging was proposed and utilized to optimize both resonator linear spring and damper properties and allowed bistable column snap-through distance. Time history analyses conducted on optimized bistable metafoundations (BMs) highlighted the favourable behaviour of these isolation systems in terms of reduction of the total seismic energy input transferred to the superstructure. Clearly, the nonlinearity that characterizes the bistable mechanism significantly reduced the total structure stiffness and resulted in a more compact locally resonant metafoundations wrt to the linear metafoundations. To further explore the dynamics and investigate the underlying properties of the bistable metafoundation (BM), the uncoupled single cell, the finite lattice and a periodic chain system of BM unit cells was investigated. Sinusoidal driven boundary inputs were applied to all systems and both time/frequency domain responses were obtained. The single cell analysis clearly showed the harmonic diffusion property. In addition, the harmonic diffusion of the antiresonance features of the resonators was also observed. The possibility of aperiodic motions were also discovered. Then, the energy transfer through finite chains was investigated by finite lattices. It has been shown that in both short and long input durations, the bistable finite lattices prevent energy transfer by snap-through motions with <10 cells. Moreover, as the input duration increases, the energy transferred to the generic cell significantly differs between the bistable and linear finite lattices; and the bistable lattice outperforms. In addition, the dispersion relations of the bistable periodic system have been computed using both analytical and numerical tools. The analytical results obtained through the harmonic balance method showed a very good agreement with the numerical approach based on the 2D FFT. Thus, the dispersion properties of the nonlinear system have been quantified.

6.8.2 Future Perspectives

With regard to the fragility assessment of industrial substructures, further analysis of other system configurations are needed, to enhance and generalise the aforementioned findings. Moreover, based on proper assumptions, the state-depend fragility curves derived from critical industrial substructures could populate Markovian transition matrices, very useful for the quantification of restoring functions.

As far as metafoudations are concerned, the complex dynamics related to bistability and wire rope hysteresis deserve further studies. More specifically, the effects of varying gravitational loads, vertical seismic excitations together with the physical realization of wire ropes and a finite lattice with the control of parameters that accurately replicate bistable nonlinear components warrants further research.

References

17-917-44 NISTGCR. (2017). Seismic analysis, design, and installation of nonstructural components and systems—Background and recommendations for future work.

Arrieta, A. F., Hagedorn, P., Erturk, A., & Inman, D. J. (2010). A piezoelectric bistable plate for nonlinear broadband energy harvesting. *Applied Physics Letters, 97*(10).

Basone, F., Wenzel, M., Bursi, O. S., & Fossetti, M. (2019a). Finite locally resonant metafoundations for the seismic protection of fuel storage tanks. *Earthquake Engineering & Structural Dynamics, 48*.

Basone, F., Wenzel, M., Bursi, O. S., & Fossetti, M. (2019). Finite locally resonant metafoundations for the seismic protection of fuel storage tanks. *Earthquake Engineering and Structural Dynamics, 48*, 232–252.

Bursi, O. S., Basone, F., & Wenzel, M. (2021a). Stochastic analysis of locally resonant linear and hysteretic metamaterials for seismic isolation of process equipment. *Journal of Sound and Vibration, 510*, 116263.

Bursi, O. S., Basone, F., & Wenzel, M. (2021b). Stochastic analysis of locally resonant linear and hysteretic metamaterials for seismic isolation of process equipment. *Journal of Sound and Vibration, 510*. https://doi.org/10.1016/j.jsv.2021.116263

Carta, G., Movchan, A. B., Argani, L. P., & Bursi, O. S. (2016). Quasi-periodicity and multi-scale resonators for the reduction of seismic vibrations in fluid-solid systems. *International Journal of Engineering Science, 109*.

Cazottes, P., Fernandes, A., Pouget, J., & Hafez, M. (2009, September). Bistable buckled beam: Modeling of actuating force and experimental validations. *Journal of Mechanical Design, 131*(10), 101001. ISSN 1050-0472. https://doi.org/10.1115/1.3179003

Celli, P., Yousefzadeh, B., Daraio, C., & Gonella, S. (2019). Bandgap widening by disorder in rainbow metamaterials. *Applied Physics Letters, 114*(9).

De Biasio, M., Grange, S., Dufour, F., Allain, F., & Petre-Lazar, I. (2015). Intensity measures for probabilistic assessment of non-structural components acceleration demand. *Earthquake Engineering & Structural Dynamics*. https://doi.org/10.1002/eqe.2582

Dubnau, D., & Losick, R. (2006). Bistability in bacteria. *Molecular Microbiology, 61*(3), 564–572.

Durstewitz, D., & Seamans, J. K. (2006). Beyond bistability: Biophysics and temporal dynamics of working memory. *Neuroscience, 139*(1), 119–133.

European Committee for Standardization. (2004a). Design of structures for earthquake resistance—Part 1: General rules, seismic actions and rules for buildings. Eurocode 8-1.

European Committee for Standardization. (2004b). Design of structures for earthquake resistance—Part 1: General rules, seismic actions and rules for buildings. Eurocode 8-1, CEN/TC 250, Brussels.

European Committee for Standardization. (2005). Design of steel structures. Part 1-1: General rules and rules for buildings. Eurocode 3-1-1.

Fajfar, P., Vidic, T., & Fischinger, M. (1990). A measure of earthquake motion capacity to damage medium-period structures. *Soil Dynamics and Earthquake Engineering, 9*(5), 236–242. https://doi.org/10.1016/S0267-7261(05)80002-8

Franchini, A., Bursi, O. S., Basone, F., & Sun, F. (2020). Finite locally resonant metafoundations for the protection of slender storage tanks against vertical ground accelerations. *Smart Materials and Structures, 29*, 5. https://doi.org/10.1088/1361-665X/ab7e1d

Frazier, M. J., & Kochmann, D. M. (2017). Band gap transmission in periodic bistable mechanical systems. *Journal of Sound and Vibration, 388*, 315–326.

Gammaitoni, L., Marchesoni, F., Menichella-Saetta, E., & Santucci, S. (1989). Stochastic resonance in bistable systems. *Physical Review Letters, 62*(4), 349.

Giaralis, A., & Petrini, F. (2017). Wind-induced vibration mitigation in tall buildings using the tuned mass-damper-inerter (TMDI). *Journal of Structural Engineering, 143*(9), 04017127.

Guner, T., Bursi, O. S., & Broccardo, M. (2023). Seismic vibration mitigation of steel storage tanks by metafoundations endowed with linear and bistable columns. *Bulletin of Earthquake Engineering*, 1–26.

Guner, T., Bursi, O. S., Erlicher, S. (2022). Optimization and performance of metafoundations for seismic isolation of small modular reactors. *Computer-Aided Civil and Infrastructure Engineering*.

Habib, G., & Romeo, F. (2017). The tuned bistable nonlinear energy sink. *Nonlinear Dynamics, 89*(1), 179–196.

Hariri-Ardebili, M. A., & Saouma, V. E. (2016). Probabilistic seismic demand model and optimal intensity measure for concrete dams. *Structural Safety, 59*, 67–85. https://doi.org/10.1016/j.strusafe.2015.12.001

Harne, R. L., & Wang, K. W. (2013). A review of the recent research on vibration energy harvesting via bistable systems. *Smart Materials and Structures, 22*(2), 023001.

Hopf, F. A., Kaplan, D. L., Gibbs, H. M., & Shoemaker, R. L. (1982). Bifurcations to chaos in optical bistability. *Physical Review A, 25*(4), 2172.

Hussein, M. I., & Frazier, M. J. (2013). Metadamping: An emergent phenomenon in dissipative metamaterials. *Journal of Sound and Vibration, 332*(20), 4767–4774.

Hussein, M. I., Leamy, M. J., & Ruzzene, M. (2014). Dynamics of phononic materials and structures: Historical origins, recent progress, and future outlook. *Applied Mechanics Reviews, 66*. https://doi.org/10.1115/1.4026911

Iervolino, I., Giorgio, M., & Chioccarelli, E. (2015b). Markovian modeling of seismic damage accumulation. *Earthquake Engineering & Structural Dynamics, 45*, 441–461. https://doi.org/10.1002/eqe.2668.

Jeffers, A. E., Plaut, R. H., & Virgin, L. N. (2008). Vibration isolation using buckled or pre-bent columns—part 2: three-dimensional motions of horizontal rigid plate. *Journal of Sound and Vibration, 310*.

Knapen, T., Brascamp, J., Pearson, J., van Ee, R., & Blake, R. (2011). The role of frontal and parietal brain areas in bistable perception. *Journal of Neuroscience, 31*(28), 10293–10301.

Laura Eads, Eduardo Miranda, Helmut Krawinkler, and Dimitrios G Lignos. An efficient method for estimating the collapse risk of structures in seismic regions. *Earthquake Engineering & Structural Dynamics*, 42(1):25–41, 2013.

Liu, C., Chen, L., Lee, H. P., Yang, Y., & Zhang, X. (2022). A review of the inerter and inerter-based vibration isolation: Theory, devices, and applications. *Journal of the Franklin Institute*.

Lorenz, E. (2000). The butterfly effect. *World Scientific Series on Nonlinear Science Series A, 39*, 91–94.

Ma, R., Bi, K., & Hao, H. (2021). Inerter-based structural vibration control: a state-of-the-art review. *Engineering Structures, 243*, 112655.

Maldovan, A. (2013). Sound and heat revolutions in phononics. *Nature, 503*, 209–217.

Malhotra, P. K., Wenk, T., & Wieland, M. (2000). Simple procedure for seismic analysis of liquid-storage tanks. *Structural Engineering International, 10*(3), 197–201.

Marian, L., & Giaralis, A. (2017). Optimal design of a novel tuned mass-damper-inerter (TMDI) passive vibration control configuration for stochastically support-excited structural systems. *Probabilistic Engineering Mechanics, 38*, 156–164.

Miniaci, M., Krushynska, A., Bosia, F., & Pugno, N. M. (2016). Large scale mechanical metamaterials as seismic shields. *New Journal of Physics, 18*(8), 083041.

Mitseas, I. P., & Beer, M. (2019). Modal decomposition method for response spectrum based analysis of nonlinear and non-classically damped systems. *Mechanical Systems and Signal Processing, 131*, 469–485.

Nardin, C., Bursi, O. S., Paolacci, F., Pavese, A., & Quinci, G. (2022). Experimental performance of a multi-storey braced frame structure with non-structural industrial components subjected to synthetic ground motions. *Earthquake Engineering & Structural Dynamics, 51*(9), 2113–2136. https://doi.org/10.1002/eqe.3656

Palermo, A., Krödel, S., Marzani, A., & Daraio, C. (2016). Engineered metabarrier as shield from seismic surface waves. *Scientific Reports, 6*(1), 1–10.

Phan, H. N., Paolacci, F., Di Filippo, R., & Bursi, O. S. (2020). Seismic vulnerability of above-ground storage tanks with unanchored support conditions for NA-tech risks based on Gaussian process regression. *Bulletin of Earthquake Engineering, 18*, 6883–6906.

Plaut, R. H., Favor, H. M., Jeffers, A. E., & Virgin, L. N. (2008). Vibration isolation using buckled or pre-bent columns—Part 1: Two-dimensional motions of horizontal rigid bar. *Journal of Sound and Vibration, 310*.

Quinci, G., Nardin, C., Paolacci, F., & Bursi, O. S. (2023). Modelling issues in seismic risk analysis of non-structural components located on industrial plant subsystems. *Bulletin Earthquake Engineering, 02*. https://doi.org/10.1002/eqe.1132

Reggio, A., Angelis, M. D. (2015). Optimal energy-based seismic design of non-conventional tuned mass damper (TMD) implemented via inter-story isolation. *Earthquake Engineering & Structural Dynamics, 44*.

Rezaeian, S., & Der Kiureghian, A. (2010). Simulation of synthetic ground motions for specified earthquake and site characteristics. *Earthquake Engineering & Structural Dynamics*. https://doi.org/10.1002/eqe.997

Rossi, L., Stupazzini, M., Parisi, D., Holtschoppen, B., Ruggieri, G., & Butenweg, C. (2020). Empirical fragility functions and loss curves for Italian business facilities based on the 2012 Emilia-Romagna earthquake official database. *Bulletin of Earthquake Engineering, 18*, 1693–1721.

Rothos, V. M., & Vakakis, A. F. (2009). Dynamic interactions of traveling waves propagating in a linear chain with an local essentially nonlinear attachment. *Wave Motion, 46*(3), 174–188.

Saif, M. T. A. (2000). On a tunable bistable mems-theory and experiment. *Journal of Microelectromechanical Systems, 9*(2), 157–170.

Salandra, V., Wenzel, M., Bursi, O. S., Carta, G., & A. B. Movchan. Conception of a 3d metamaterial-based foundation for static and seismic protection of fuel storage tanks. *Frontiers in Materials, 4*.

Smith, M. C. (2002). Synthesis of mechanical networks: the inerter. *IEEE Transactions on Automatic Control, 47*(10), 1648–1662.

Specker, T., Buchholz, M., & Dietmayer, K. (2014). A new approach of dynamic friction modelling for simulation and observation. *IFAC Proceedings Volumes, 47*(3), 4523–4528.

Vathi, M., Karamanos, S., Kapogiannis, I., & Spiliopoulos, K. (2017, July). Performance criteria for liquid storage tanks and piping systems subjected to seismic loading. *Journal of Pressure Vessel Technology, 139*. https://doi.org/10.1115/1.4036916

Wenzel, M., Bursi, O. S., & Antoniadis, I. (2020). Optimal finite locally resonant metafoundations enhanced with nonlinear negative stiffness elements for seismic protection of large storage tanks. *Journal of Sound and Vibration, 483*. https://doi.org/10.1016/j.jsv.2020.115488

MIX
Papier aus verantwortungsvollen Quellen
Paper from responsible sources
FSC® C105338

If you have any concerns about our products,
you can contact us on
ProductSafety@springernature.com

In case Publisher is established outside the EU,
the EU authorized representative is:
Springer Nature Customer Service Center GmbH
Europaplatz 3, 69115 Heidelberg, Germany

Printed by Libri Plureos GmbH
in Hamburg, Germany